"Compassionate Digital Innovation reveals that because AI companions are emotional infrastructure shaped by loneliness, power, and profit, the difference between care and extraction lies not just in good intentions but also in how we design, govern, and take responsibility for them. Its call for dignity and accountability in design resonates strongly with our commitment to building AI that genuinely supports women's wellbeing."

Sreyna Rath, *Founder and CEO of Jaimee AI*

"Compassionate Digital Innovation offers a principled and urgently needed framework for designing AI systems that prioritise child safety, dignity, and long-term development. By recognising that emotionally responsive AI and digital infrastructures are not merely technical tools but developmental and societal forces, Ciriello makes a compelling case for embedding compassion, accountability, and evidence-based safeguards into innovation before harm scales."

John Zoltner, *Founder and CEO of AIChildSafety.org*

"Within a world that evokes us to question what it means to be human, Compassionate Digital Innovation creates a safe space for heartfelt philosophical inquiry. Each page leads us to engage not only our intellect but our imagination, inviting us to consider new versions of existence where technology empowers the individual, not the powers that be. This text is an essential companion for anyone who desires to veer technological trajectories in directions of liberation."

Kathryn Carter, *Poet and Editor of Les Lettres*

"Compassionate Digital Innovation is a wakeup call for our generation. It reminds us that if we allow technology to shape our children without wisdom and discernment, we risk losing the very humanity we are called to protect. This book challenges us in our leadership to guide AI in ways that strengthen families and real community!"

Rev. Noddy Sharma, *Founder of Abel Movement and AI Future Generations Think Tank*

Compassionate Digital Innovation

Compassionate Digital Innovation examines the ethical tensions at the heart of the digital innovations that shape contemporary life, from artificial intelligence and blockchain to social media platforms. Rather than treating harm as an accidental by-product of progress, this book argues that recurring dilemmas are embedded in the structural logics of digital infrastructures.

Through a pluralistic conceptual framework, it shows how amplification, personalisation, persistence, automation, and simulation generate predictable tensions at scale. Drawing on moral philosophy and empirical research, this book proposes compassion as a constitutional orientation for navigating these tensions. Compassion is framed not as sentimentality, but as a disciplined commitment to reduce avoidable suffering and redistribute responsibility in proportion to power. Across social media, Web3 governance, and AI companionship, readers encounter concrete analyses of both promise and peril. This book moves beyond appeals to "personal responsibility" to envision organised compassion through resistance to extractive models, regulatory reform, and the construction of public-interest alternatives.

For managers, technologists, policymakers, researchers, students, and engaged citizens, this book offers a rigorous yet accessible framework for redesigning digital innovation in service of dignity, justice, and the common good.

Raffaele F. Ciriello is a scholar of compassionate digital innovation at the University of Sydney, Australia. His research advances a critical dialectical approach, focusing on ethical tensions in the design, governance, and use of emerging technologies. He is a frequent contributor to leading journals (such as the *AIS Basket*, *Nature Machine Intelligence*, and *The Lancet*) and media outlets (such as *The Economist*, *Financial Times*, and *60 Minutes*).

Compassionate Digital Innovation

Navigating the Ethical Dilemmas of Sociotechnical Change

Raffaele F. Ciriello

Routledge
Taylor & Francis Group
LONDON AND NEW YORK

First published 2027
by Routledge
4 Park Square, Milton Park, Abingdon, Oxon OX14 4RN

and by Routledge
605 Third Avenue, New York, NY 10158

Routledge is an imprint of the Taylor & Francis Group, an informa business

British Library Cataloguing-in-Publication Data
A catalogue record for this book is available from the British Library

ISBN: 9781041014539 (hbk)
ISBN: 9781041014546 (pbk)
ISBN: 9781003614814 (ebk)

DOI: 10.4324/9781003614814

Typeset in Times New Roman
by codeMantra

Contents

Figures

Tables

Acknowledgements

Ever since first grade I wanted to become an author. Over the years I have written many academic articles, but this book makes the aspiration feel real for the first time. It allowed me to think in longer arcs, to connect arguments across chapters, and to say things that do not easily fit within the disciplined constraints of journal formats. I am deeply grateful for the space to do so. Even so, this book is anything but a solitary achievement.

First and foremost, I remain forever indebted to my loving wife, Mengia. My introverted inner child once imagined that becoming an author would involve retreating to a secluded cabin by a quiet lake, sustained by coffee, ramen, and solitude, armed with nothing but a typewriter. My reality looked quite different. With three little children and a full-time academic role, writing this book meant negotiating time in early mornings, late nights, foregone weekends, and stolen hours between responsibilities. Without a rockstar mother and project manager holding the fort (and occasionally holding me accountable), none of this would have been possible. Her steadiness, humour, and patience made intellectual ambition compatible with family life.

I am sincerely grateful to my editor, Kendrick Loo, whose thoughtful feedback and steady encouragement strengthened both the manuscript and my confidence in it. A good editor improves sentences. A great one improves the clarity of the author's thinking. I was fortunate to have the latter.

I thank the University of Sydney for offering an environment where critical, unconventional scholarship can breathe, and for supporting the publication of this book.

Finally, I want to thank the many colleagues, students, and friends who enriched my thinking through countless conversations about compassionate digital innovation. Some challenged assumptions, some sharpened arguments, some offered a much-needed reality check, and others reminded me that ideas only matter when they travel beyond the page. If this book carries coherence, it is because it was refined in dialogue. Whatever insights endure here are indebted to these relationships. Any remaining shortcomings are, of course, my own. And may any typos you find sprinkle joy in your day.

Synopsis

Digital technologies increasingly shape how we communicate, work, govern, love, and remember. Social media organises public discourse. Blockchain promises decentralised trust. AI companions simulate intimacy and emotional support. Each wave arrives with aspirations of connection and empowerment, yet repeatedly generates harm: polarisation, surveillance, exploitation, dependency, and structural inequality. Why do these patterns persist, and how might we redirect them?

Compassionate Digital Innovation argues that harm is not simply the product of malicious engineers, greedy firms, or careless users, nor reducible to technical flaws or regulatory inaction. It arises from recurring structural logics embedded in digital infrastructures: amplification, personalisation, persistence, automation, and simulation. These affordances scale human expression and coordination in unprecedented ways. When embedded in profit-driven and power-concentrated systems, they predictably generate ethical tensions. Technologies that amplify connection also amplify outrage. Personalisation fragments shared realities. Automation obscures responsibility. Simulation can comfort yet deceive.

This book develops a framework for navigating these tensions without lapsing into technological determinism or naïve optimism. Digital innovation consistently produces enduring dilemmas: progress versus preservation, intention versus outcome, past versus present versus future, autonomy versus coordination, and unity versus diversity. These cannot be optimised away; they are constitutive of sociotechnical systems at scale. What is needed is not a technocratic fix but an orienting principle capable of holding competing values in productive tension.

This book proposes compassion as that principle. Drawing on moral philosophy and social psychology, compassion is framed not as sentimentality but as a disciplined commitment to minimise avoidable suffering and extend moral concern beyond narrow in-groups. Distinguished from empathy's selective spotlight, it functions as a scalable meta-ethical anchor grounded in the dual imperative to hurt nobody and to help where possible.

A central claim of this book is that virtue alone is insufficient. Digital harms scale through infrastructures. Compassion must therefore scale as well. The final chapters examine how resistance, removal, and replacement can institutionalise compassion within governance. Revisiting Luddism, this book distinguishes reactionary technophobia from principled resistance to exploitative deployment. Organised compassion becomes both critique and construction.

This book concludes by turning to personal and professional commitments. It challenges students, engineers, managers, and policymakers to reconsider where they invest their skills and ambition. Career choices, institutional prestige, and resource allocation are ethical acts. Organised compassion requires not only new rules but also new norms of aspiration.

At its core, *Compassionate Digital Innovation* offers both diagnosis and direction. By grounding innovation in disciplined compassion, societies can move from reactive harm mitigation toward proactive, dignity-centred design. The future of digital technology is not predetermined. It depends on how we organise innovation today.

1 Digital Innovation as a Double-Edged Sword

Digital innovation is now a defining force of contemporary life. Social interaction, political deliberation, economic exchange, and even emotional support are mediated through platforms and algorithms. What once seemed like tools now operate as infrastructures that organise collective life. The dominant narrative is that, despite occasional harms, digital technologies produce net benefits.[1] From this perspective, innovation is presumed progressive, and ethics becomes a matter of managing unfortunate side effects.

This book begins from a different premise. Digital innovation is not unambiguously progressive, nor are its harms accidental deviations from a benevolent trajectory. It is structurally double-edged. The infrastructures that connect also divide. The systems that empower also exploit. The architectures that promise decentralisation can consolidate power in new forms. Whether digital innovation produces a net positive is not a settled empirical fact, but a moral and political judgement shaped by perspective, position, and power.[1]

Even the metaphor of a "double-edged sword" is, itself, double-edged. It implies symmetry: equal potential for good or bad, depending on how the sword is wielded. This framing can deflect responsibility for systemic harm onto individuals. If harm occurs, users are often blamed for having misused the technology, engineers are accused of normless tunnel vision, and corporate managers are cast as greedy and exploitative.

What remains hidden from this view are asymmetries of design, incentive, and control. When platforms are optimised for engagement, when profit depends on data extraction, and when amplification is frictionless because volatility is profitable, harms are not simply isolated lapses. They are foreseeable, recurring outcomes of institutional arrangements. The ambiguity of digital innovation is therefore structural, not merely moral.

Three assumptions embedded in mainstream discourse deserve reconsideration. First, that digital innovation produces aggregate welfare gains sufficient to justify its expansion. Second, that ethical evaluation can be reduced to a consequentialist tally of benefits and harms. Third, that emerging harms can be corrected through incremental corporate reform and regulatory adjustment.[1]

DOI: 10.4324/9781003614814-1

Each assumption is incomplete. Aggregate gains often conceal distributional asymmetries. Platform economies have generated enormous shareholder value while exposing gig workers to precarity and users to behavioural manipulation.[2] Social media has lowered barriers to expression while amplifying polarisation and misinformation. Blockchain technologies have enabled new forms of coordination while imposing environmental costs and fuelling speculative excess. AI systems promise efficiency and access while embedding biases that disproportionately burden marginalised communities. When benefits accrue to a privileged few and burdens diffuse widely across vulnerable populations, aggregate welfare provides an ethically thin account.

Ethical evaluation cannot be reduced to outcome calculation when rights, dignity, and justice are at stake. An algorithm that maximises engagement may increase overall utility while degrading public discourse. A predictive model may improve average accuracy while systematically disadvantaging minority groups. Consequentialist reasoning remains important, but it cannot exhaust the moral field when fairness, voice, and long-term institutional integrity are at stake.

Regulatory patchwork also struggles to keep pace with technologies designed for global scale and rapid iteration. Content moderation reforms lag behind new forms of manipulation. Privacy regulations are constrained by jurisdictional boundaries. AI governance frameworks struggle to address fast-moving capabilities. Incremental adjustment can mitigate harm, but it rarely challenges the incentive structures that generate it.

At the same time, digital innovation is unlikely to recede. Contemporary societies are deeply dependent on networked communication, algorithmic coordination, and data-driven systems. A wholesale retreat from digital technologies is neither realistic nor desirable. This is not an argument for resignation or manufactured consent. It is a recognition that digital infrastructures now constitute basic conditions of social life. The question is not whether innovation should continue, but how it can be redirected toward the common good.

By the common good, I do not mean the maximisation of aggregate utility or shareholder value. I refer to the shared conditions that enable individuals and communities to flourish across social, economic, and environmental dimensions.[1,3,4] This includes the well-being of all stakeholders – individuals, organisations, communities, and the natural environment – whose interests are directly or indirectly affected by digital technologies. Some interests are immediate, such as those of users or employees. Others are diffuse, such as ecological impacts or long-term social disruption.

Aligning digital innovation with the common good is not straightforward. Protecting some may impose burdens on others. Content moderators absorb psychological harm to shield users from abuse. Intervening against online harassment can be risky and emotionally draining. Systems that optimise efficiency may advantage some groups while entrenching inequality for others. If such dilemmas are ignored, the benefits of digital innovation can erode into systemic harm. Serving the common good therefore requires protecting dignity, distributing benefits and burdens fairly, and preserving shared resources for future generations.

The ethical complexity of digital innovation stems from three interrelated characteristics: scale, opacity, and concentration. Scale transforms local actions into global consequences. A design change can alter the information diet of millions. A biased model can shape access to credit or employment at scale. Opacity obscures causal chains. Proprietary algorithms and distributed systems complicate responsibility, enabling plausible deniability even when harm is evident and severe.[5,6] Concentration compounds these dynamics. Despite rhetoric of decentralisation, much of the digital infrastructure that governs communication, value exchange, and even emotional life is controlled by a small number of corporations.[7,8] Decisions affecting democratic participation and relational intimacy are often made in boardrooms rather than accountable public institutions.

Against this backdrop, compassion emerges as both necessary and contested. It is necessary because digital technologies increasingly mediate encounters with vulnerability – of users experiencing loneliness, of workers moderating traumatic content, and of communities targeted by discrimination. It is contested because it should not be reduced to branding or instrumentalised for reputational gain while underlying incentives remain unchanged. The challenge is to reclaim compassion as a principled orientation rather than a marketing slogan.

Digital innovation refers here to the process by which new technologies are designed, governed, and used to drive economic, social, and environmental change.[1] These innovations promise efficiency, scalability, and access to resources that were previously unattainable. However, this progress has often been accompanied by societal harms, including privacy violations, social disenfranchisement, environmental degradation, and economic inequality.[1] As the reach and influence of digital technologies grow, so too does the need for an ethical framework that can guide their responsible design, governance, and use.

Compassion, from the Latin "to suffer with", goes beyond empathy.[9] It entails recognising suffering and acting to alleviate it.[1] In the context of digital innovation, compassion demands deliberate attention to how technologies affect those most vulnerable to harm. It calls for design, governance, and use that prioritise social well-being over profit maximisation.

The core claim of this book is threefold. First, digital innovation is structurally ambiguous. Its affordances generate dual potentials that, when scaled and institutionalised under particular incentives, crystallise into recurring ethical dilemmas. Second, prevailing evaluative frameworks – especially narrow consequentialism – are insufficient to navigate these dilemmas under conditions of scale, opacity, and institutional concentration. Third, compassion, properly understood, can function as a meta-ethical orientation capable of anchoring pluralism and guiding technology design, governance, and use toward the common good. Compassion, in this sense, is not sentimentalism but disciplined attention to preventable suffering in environments designed to obscure it.[1] On this foundation, this book advances a programme of institutional redesign grounded in democratic oversight and needs-based justice.

Compassion, as developed in this book, does not deny trade-offs. Rather, it provides a stable reference point in environments of irreducible tension. It resists technocratic optimisation, which reduces ethics to metrics, and relativism, which dissolves accountability into preference. By orienting attention toward those who bear disproportionate burdens, compassion constrains extraction and guides institutional reform.

This book unfolds in three parts: concepts, critique, and change. This trifold structure reflects a conviction that diagnosis without prescription risks cynicism, while prescription without diagnosis risks blind action for action's sake.

The first part develops the **concepts** required to understand why digital innovation predictably generates ethical dilemmas. *Chapter 2* introduces the notion of affordance ambiguity. Digital technologies offer action possibilities that are inherently double-edged, becoming beneficial or harmful depending on how they are actualised at scale under asymmetrical power. These ambiguities stabilise into recurring dilemmas, discussed in Chapter 3, that cannot be reduced to technical optimisation problems because they reflect deeper struggles over power, vulnerability, and legitimacy. Compassion, discussed in Chapter 4 and understood as a principled commitment to "hurt nobody" and to "help wherever possible", offers a principled way forward to navigate these dilemmas without collapsing into either rigid dogmatism or unbounded relativism.

The second part offers a **critique** of three successive waves of digital innovation: social media, Web3, and AI companions. *Chapter 5* examines social media, emblematic of Web 2.0. Celebrated for democratising speech and lowering barriers to participation, it amplified marginalised voices and enabled social mobilisation. Yet it also institutionalised volatile amplification, algorithmic personalisation, and persistent digital traces, reshaping public discourse and entrenching surveillance capitalism. Connection became a marketplace where attention is monetised, and reputations remain perpetually exposed. *Chapter 6* turns to Web3 and blockchain technologies, which emerged in response to centralised platform power. Promising decentralisation, transparency, and trust-minimised value exchange, they enabled experiments in cooperative ownership and programmable governance. At the same time, they produced speculative excess, governance capture, and environmental externalities. Power was not eliminated but reconfigured in code and token distribution. *Chapter 7* addresses AI companions, the frontier of artificial emotional intelligence. Marketed as remedies for loneliness, they offer simulated care. For some, they provide comfort. For others, they foster dependency and extract sensitive data. As automated systems enter domains of intimacy and grief, the moral stakes escalate. Together, these technologies trace a trajectory from communication to value exchange and simulated intimacy, each embodying ambiguous affordances scaled under concentrated power.

The third part advances **change**. Ethical reflection cannot remain at the level of critique. If digital technologies structure the common good, compassionate digital innovation requires transformation at multiple levels: individual orientation, collective organisation, and infrastructural design. *Chapter 8* explores how compassion can be cultivated within persons. *Chapter 9* discusses how compassion can be organised through unions, open-source communities, and civil society coalitions. Finally, *Chapter 10* envisions how compassion can be institutionalised through public digital infrastructures anchored in democratic oversight.

This book is not another manual of "corporate responsibility" nor a catalogue of regulatory proposals. It offers a theoretical framework for navigating ethical dilemmas in digital innovation and argues that without an orienting moral commitment capable of confronting that ambiguity, reforms will remain superficial. Compassion, in this account, is not a performative virtue but a constitutional orientation. It demands that the infrastructures shaping intimacy, identity, and democracy be designed, governed, and used with explicit attention to those who bear disproportionate burdens. Figure 1.1 visualises the framework, and the rest of this book unpacks it.

We are living through a vast sociotechnical transformation whose outcomes are not predetermined. Digital infrastructures remain malleable. Governance arrangements remain contested. The question is whether future waves of innovation will deepen extraction and power concentration or be redirected toward dignity, justice, and the common good. The difference will not be decided by technology alone, but by how we understand its structural ambiguities, confront its dilemmas, and embed compassion at the heart of its design, governance, and use.

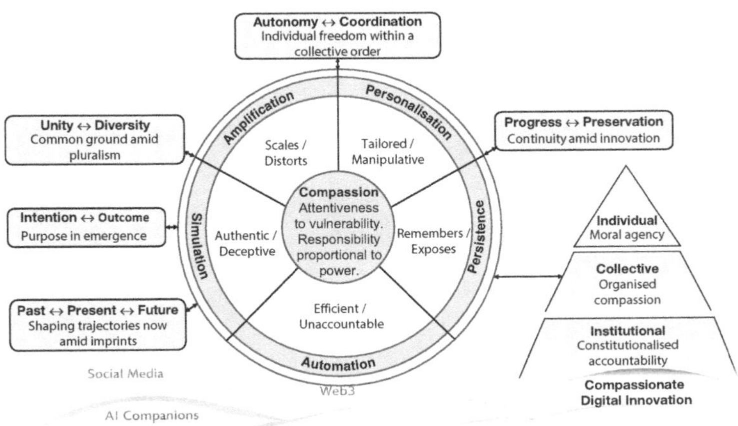

Figure 1.1 A Framework for Navigating Ethical Dilemmas in Digital Innovation.

References

1 Ciriello, R. F., Mathiassen, L., Risius, M., Cheong, M., Scheepers, H., & Vaast, E. Compassionate Digital Innovation: A Pluralistic Perspective and Research Agenda. *Information Systems Journal* **36**, 364–385, doi:10.1111/isj.70003 (2025).
2 Arsel, Z. Platform capture: a review of the state of the art of research on platforms and a research agenda. *Journal of Marketing Management*, 1–21, doi:10.1080/0267257X.2025.2549741 (2025).
3 Chomsky, N. in *Columbia University Public Lecture, 6 December 2013* (New York City, 2013).
4 Argandoña, A. The stakeholder theory and the common good. *Journal of Business Ethics* **17**, 1093–1102, doi:10.1023/A:1006075517423 (1998).
5 Birsch, D. Moral responsibility for harm caused by computer system failures. *Ethics and Information Technology* **6**, 233–245, doi:10.1007/s10676-005-5609-5 (2004).
6 Ciriello, R. F., Gal, U., Hannon, O. & Thatcher, J. Responsible social media use: how user characteristics shape the actualisation of ambiguous affordances. *European Journal of Information Systems* **34**, 799–821, doi:10.1080/0960085X.2024.2444249 (2025).
7 Ciriello, R. F., Marx, J., Cheong, M., Mueller-Bloch, C. & Mathiassen, L. Decentralized social media. *Business & Information Systems Engineering* **68**, 219–234, doi:10.1007/s12599-025-00952-4 (2025).
8 Ciriello, R. F. & Mindel, V. Techno-collectivism: scaling libertarian socialist governance of digital infrastructure. *IEEE Technology and Society Magazine* **44**, 68–82, doi:10.1109/MTS.2025.3605115 (2025).
9 Bloom, P. *Against Empathy: The Case for Rational Compassion*. (Random House, 2017).

Part I
Concepts

Part I
Concepts

2 Ambiguous Affordances of Digital Technologies

It is a simple but often overlooked premise: digital technologies do not produce ethical outcomes by virtue of their features alone. They matter morally because of the action possibilities they open, constrain, and channel. These possibilities – affordances – are inherently ambiguous.[1] When actualised at scale, they crystallise into recurring ethical dilemmas. Because these dilemmas are structurally produced rather than accidental, navigating them demands not just technical refinement, but moral reorientation.

The concept of *affordance*, originating in Gibson's ecological psychology and later refined in information systems research, shifts attention from tools to relationships. Affordances are goal-oriented action possibilities that arise in the relation between a technology and a user situated in context.[1] They are neither purely objective properties nor subjective projections. Instead, they are relational potentials that become consequential only when they are perceived, interpreted, and enacted. In other words, outcomes emerge through affordance actualisation rather than from features alone.[1] A "like" button does nothing until users interpret and enact it.[2] A blockchain ledger does not decentralise power until participants mobilise around its protocols.[3] A chatbot does not "care" until someone confides in it.[4] What matters ethically is not that a feature exists, but how it becomes woven into practice, sometimes in ways that exceed the foresight of any single designer, policymaker, or user.

This relational perspective clarifies why responsibility in digital innovation is difficult to locate and easy to diffuse. If outcomes emerge through distributed actualisation, effects are neither predetermined by technological features nor fully controlled by users. They result from sociotechnical processes shaped by users' goals, interpretations, and characteristics, as well as by institutional arrangements that structure possible action.[1] In platform environments, affordances are rarely offered neutrally. They are engineered and nudged through revenue models, ranking algorithms, interface defaults, moderation rules, and data infrastructures that reward some actions and penalise others.

Such a relational framing guards against two errors. The first is technological determinism, which treats harmful outcomes as inevitable. The second is

DOI: 10.4324/9781003614814-3

naïve voluntarism, which assumes users are entirely free to shape technologies as they wish.[2] In practice, affordances are actualised within institutional environments structured by incentives, norms, and governance. These environments often privilege growth, engagement, and data extraction over dignity, fairness, or social cohesion.

Affordance ambiguity, in this sense, is structural rather than accidental. Attempts to eliminate such ambiguity, though popular in mainstream research due to its preference for conceptual clarity, are futile. Digital technologies increasingly operate in morally charged domains, spanning intimacy, vulnerability, political identity, desire, and grief. In such domains, dual potentials are predictable consequences of design. Confideability in an AI companion can allow someone to vent without burdening loved ones while simultaneously generating extractive pipelines of intimate disclosure.[4] Spreadability of social media content can amplify marginalised voices yet annihilate reputations within hours.[1] A cryptocurrency token can expand financial access while funding fraud and coercion. If outcomes are produced through actualisation in institutionally governed contexts, ethical dilemmas are not peripheral to digital innovation, but embedded in its sociotechnical logic.

Five Key Affordances of Digital Innovation

Across social media, Web3 architectures, and AI companions – the three technologies examined in Chapters 5 through 7 – a recurring set of affordances is visible. Rather than cataloguing features, we can identify five structural logics that travel across technologies: amplification, personalisation, persistence, automation, and simulation. Their ethical intensity emerges when they stack. Table 2.1 provides an overview.

Amplification refers to the scaling of expression, visibility, coordination, and harm. On social media, volatile spreadability allows content to move from obscurity to global attention within hours, often detached from context. A flippant remark on Twitter/X may go viral in unexpected ways. Amplification enables mobilisation, mutual aid, and emergency coordination; it also enables harassment and reputational harm.[1] Design makes scale frictionless while consequences remain opaque. In Web3 systems, amplification operates through network effects coupled with tokenised incentives. Participation can scale into community governance and experimentation, yet extraction can scale into speculative bubbles and concentrated power.[5] In AI companions, amplification takes the form of scalable care-like interaction. Emotional responsiveness can be delivered to millions instantaneously. A relational resource once constrained by human attention becomes reproducible at industrial scale, and the moral stakes scale accordingly.

Personalisation captures the adaptive tailoring of digital experiences. On social media, algorithmic targetability curates feeds according to inferred preferences. Algorithmic content curation can foster fine-grained exploration and

Table 2.1 Five Key Affordances of Digital Innovation

Affordance	*Definition*	*Examples*
Amplification	The scaling of expression and visibility across large populations. Dual in potential because mechanisms that magnify voice and mobilisation also magnify error, outrage, and harm.	**Social media:** Viral spread enables rapid mobilisation and mutual aid, yet accelerates harassment and reputational damage detached from context. **Web3:** Network effects scale governance participation, yet inflate speculation and concentrate token-based power. **AI companions:** Scalable emotional responsiveness extends care-like interaction to millions, yet commodifies intimacy and fosters dependency.
Personalisation	The adaptive tailoring of content, interaction, or rules to inferred preferences. Dual in potential because relevance depends on sensitive data that can empower users or render them manipulable.	**Social media:** Algorithmic curation supports discovery and learning, yet narrows horizons and exploits vulnerabilities for engagement. **Web3:** Tokenised governance enables customised participation, yet creates exclusionary enclaves and personalised extraction. **AI companions:** Adaptive tone and memory enable emotional attunement, yet reinforce unhealthy scripts and deepen dependence.
Persistence	The durability and retrievability of digital traces over time. Dual in potential because lasting records enable accountability and collective memory while entrenching exposure and limiting forgiveness.	**Social media:** Archivable posts preserve memory and support accountability, yet tether identities to permanent past mistakes. **Web3:** Immutable ledgers enhance transparency, yet radicalise exposure through irreversible histories. **AI companions:** Conversational memory sustains continuity, yet accumulates intimate disclosures as enduring corporate data assets.
Automation	The delegation of tasks or decisions to algorithmic systems. Dual in potential because efficiency and scale displace contextual judgement and diffuse accountability.	**Social media:** Algorithmic ranking and moderation coordinate content at scale, yet obscure responsibility and reproduce bias. **Web3:** Smart contracts execute rule-based allocation, yet encode early power asymmetries into automated distribution. **AI companions:** Synthetic reassurance provides immediacy, yet risks moral misalignment in moments requiring discernment.
Simulation	The production of human-like presence or legitimacy through digital mimesis. Dual in potential because felt authenticity can distort reciprocity and ethical judgement.	**Social media:** Curated authenticity and synthetic media expand expression, yet foster deception and identity manipulation. **Web3:** Rhetoric of decentralisation signals autonomy, yet masks informal authority and elite capture. **AI companions:** Expressive malleability offers companionship, yet performs care without moral agency.

learning while narrowing horizons, entrenching echo chambers, and enabling manipulation through inferred vulnerabilities.[1] In AI companions, personalisation deepens into evolving interactability. Users customise tone and persona; systems remember disclosures and adapt to micro-signals of mood. This may support those navigating loneliness or social anxiety, yet it can reinforce unhealthy scripts, deepen dependency, and normalise extractive intimacy. In Web3, tokenised architectures enable tailored governance and incentive schemes while generating exclusionary enclaves and personalised extraction. Personalisation always raises a question: who benefits from intimate knowledge of the user, and to what end? Too often, under the advertising-based business models that dominate today's social media, the user effectively becomes the product whose data and attention are being sold to advertisers.

Persistence denotes the endurance of digital traces. On social media, searchable archives and the afterlife of content preserve collective memory and enable accountability, yet they can impose permanent punishment for transient misjudgements. Identity becomes tethered to durable histories. Public blockchains radicalise persistence through irreversibility, whereby transactions and reputations are publicly legible and, in principle, unalterable. This legibility can enable transparency while entrenching exposure and stratification. In AI companions, persistence takes the form of conversational memory, that is, vast archives of human intimacy stored on corporate servers. Confessions accumulate, creating durable information assets governed in ways often opaque to users and regulators.

Automation refers to delegated action. Systems execute, recommend, moderate, and distribute without continuous human judgement. On social media, algorithmic ranking and automated enforcement shape visibility and sanction. In Web3, smart contracts and reward mechanisms allocate value according to predefined rules. Steem's model of "getting paid for figuring out who should get paid" illustrates a hybrid logic: human evaluation feeds automated distribution, yet stake-weighting skews outcomes toward the already powerful.[5] In AI companions, automation appears as synthetic empathy, expressed as automated reassurance and affective mirroring.[4] In ordinary distress, such delegation can offer immediacy. In crisis, the absence of moral discernment can be catastrophic. Automation enables efficiency and coordination while risking moral outsourcing. Such decisions become morally consequential when they lack a clearly accountable agent.

Simulation captures the production of human-like mimesis through social presence, perceived responsiveness, and relational cues. It is most visible in AI companions, where expressive malleability creates the outward shape of care. Simulation enables experimentation and felt presence, yet it can foster dependency, distort reciprocity, and commodify intimacy. The system "performs compassion" without possessing genuine moral discernment, agency, or accountability.[4] Simulation also shadows other technologies. Social media is saturated with curated authenticity and increasingly flooded with 'AI slop', that is, synthetic low-quality, high-volume, and often misleading multimedia content generated rapidly by AI tools with little to no human oversight.[6] Web3 ecosystems often deploy rhetoric of trustlessness and decentralised virtue that can simulate ethical purity while masking new concentrations of power. Simulation then recruits users into deceptive relationship patterns without relational accountability.

Each affordance generates dual potentials. Amplification mobilises voice while facilitating harassment. Personalisation enhances relevance while enabling exploitation. Persistence secures accountability while entrenching exposure. Automation delivers scalability while displacing judgement. Simulation offers comfort while industrialising intimacy and distorting ethical perception. These dual potentials do not arise because some users are malicious and others benevolent. They arise because affordances are double-edged by nature. Ethical dilemmas emerge in the mainstream. When amplification combines with personalisation and automation, outrage economies flourish. When persistence stacks with simulation and automation, intimate disclosures become durable data assets repackaged and sold for profit. When all five logics converge, technologies shift from tools to infrastructures shaping identity, intimacy, and democracy.

This implication is foundational for the rest of the book. Ethical conflict in digital innovation is not an anomaly. It is the predictable outcome of ambiguous affordances actualised at scale within institutional contexts structured by misaligned incentives and concentrated power. The critical question is thus not whether dual potentials exist, but how they can be navigated.

The later chapters will make this mechanism concrete in three domains that represent successive waves of digital transformation. Chapter 5 examines how social media's emancipatory promise – democratising speech and enabling mobilisation – is inseparable from volatile amplification, algorithmic personalisation, and persistent traces that intensify polarisation, misinformation, surveillance capitalism, and reputational exposure. Chapter 6 analyses Web3's promise of autonomy – permissionless entry, forkability, immutability, token governance – alongside coordination failure, informal authority, gateway recentralisation, elite capture, and tokenomics that fuse governance rights with speculative assets. Chapter 7 turns to the intimate sphere, where AI companions offer accessible simulated care while creating pipelines of relational disclosure, dependency, and concentrated control over emotional data – including, in extreme cases, lethal mismatches between statistical fluency and moral discernment.

For now, the conceptual lesson is already visible. Affordances generate dual potentials. Dual potentials stabilise into recurring ethical dilemmas. Navigating them requires more than optimisation. It requires a moral orientation capable of holding ambiguity without collapsing into cynicism or actionism. Compassion, developed in Chapter 4, is proposed as that orientation – not because it eliminates ambiguity, but because it disciplines attention to suffering within systems that make harm scalable, personalisable, persistent, automated, and simulated.

Ethical complexity is not a bug in digital innovation, but its defining feature.

How Affordance Ambiguity Generates Ethical Tensions across Key Technologies

Across this book, three technologies – social media, Web3 infrastructures, and AI companions – serve to examine successive waves of digital innovation. They are not treated as discrete case studies but as different surface expressions of the same underlying sociotechnical mechanism: ambiguous affordances actualised at scale under concentrated power. Each technology arrives with emancipatory promise. Each reorganises visibility, agency, and coordination in ways that feel initially liberating. And each generates harms that are not incidental "side effects" but patterned outcomes of how particular affordance logics stack and stabilise at the systemic level. The purpose of this section is therefore neither to claim that these technologies are identical nor to merely enumerate them, but to show how a small set of recurring affordances – amplification, personalisation, persistence, automation, and simulation – repeatedly crystallise into the ethical dilemmas introduced in Chapter 3.

Chapter 5 shows that social media's moral volatility is not primarily a product of deviant use, but of mainstream design and structural conditions. The familiar promise of democratised participation depends on affordances that also make public life brittle. Amplification ensures that ordinary expression can travel beyond its intended audience at extraordinary speed and scale. Personalisation ensures that content is routed toward those most likely to react strongly, because engagement is both measurable and profitable. Persistence means that once something becomes visible, it can remain searchable, indexable, and endlessly reuploaded and reshared, often detached from its original context and intent.

The ethical dilemma here is not merely that "bad people display bad manners online". It is that platforms convert local speech into a scalable consequence. A single utterance, once amplified and recontextualised, can become a global spectacle. This does not require bad faith or malice. It requires only routine participation in an environment where amplification is frictionless, personalisation is engagement-optimised, and persistence grants content an afterlife that exceeds the temporal horizon of the speaker.

Under these conditions, intention is routinely eclipsed by outcome. Moral judgement becomes decoupled from the interpretive context in which speech was produced, because the platform's architecture prioritises virality over situated meaning. Social media thereby produces a distinctive form of moral exposure: reputations become perpetually vulnerable to collective escalation, while responsibility becomes diffuse because harms are generated by distributed action within an engineered attention economy.

Chapter 6 shows that Web3's vocabulary differs – decentralisation, trust-minimisation, permissionless entry – but its ethical tensions are shaped by the same structural logics. The Steem case and related tokenised governance experiments illustrate how persistence, automation, and personalisation can generate both empowerment and oligarchic concentration. Persistence, via public ledgers, renders transactions, balances, and reputation signals durable and legible, producing transparency but also enduring exposure. Automation, via smart contracts and code-based incentive mechanisms, promises fairness through rule-based distribution, but it also converts governance into a system that can be gamed, captured, or tilted by initial conditions. Personalisation, via interfaces and delegation mechanisms, expands agency by allowing users to curate communities and outsource voting, while simultaneously enabling informal blocs and coordinated capture.

The lesson is not that blockchain failed or that an otherwise pure design was corrupted. It is that tokenised architectures fuse governance with capital. Once automated reward pools are weighted by stake, influence accumulates precisely where it is already concentrated. Persistent public balances become instruments of stratification and surveillance. Delegation consolidates power through convenience: coordination becomes easier for those with resources, networks, and technical expertise. What begins as a vision of decentralised coordination can therefore produce paradoxical effects: participation intertwined with concentrated control, transparency intertwined with surveillance, and autonomy intertwined with elite capture. These outcomes are not aberrations but predictable trajectories of affordance actualisation under incentives that reward cumulative advantage without commensurate mechanisms to safeguard collective legitimacy.

Chapter 7 pushes this critique into a domain where the moral stakes become more intimate and, therefore, more severe. AI companions are marketed as supportive, adaptive, and always available. Their appeal is grounded in a particular stacking of affordance logics: amplification, personalisation, persistence, automation, and simulation converge around psychological vulnerability. Accessibility scales companionship beyond the scarcity of human attention. Adaptability allows tone, persona, and relational script to be shaped around the user. Confideability lowers the social cost of disclosure by offering interaction without fear of burdening others or being judged. Conversational memory preserves disclosures and relational continuity over time. Automated affective response performs reassurance and counsel in moments of distress. Simulation produces felt presence through human-like cues without reciprocal moral agency.

The dual potential is therefore built into the very foundation of these tools. AI companions can provide comfort, rehearsal space, and emotional continuity. They can also intensify dependency, blur boundaries between human and machine, and create pipelines of intimate data extraction. Most importantly, when automated systems enter domains of loneliness, trauma, and grief, failure is not merely an inconvenience. It can become life-shaping. The resulting "intimacy economy" is the structural expression of stacked affordances operating in emotionally loaded contexts. Systems are designed to elicit disclosure, remember it, and respond in ways that feel caring, while remaining embedded in commercial infrastructures where data is valuable and governance is opaque.

Taken together, these three technologies show that affordance ambiguity is not an abstract interpretive overlay. It is the organising logic that unifies the empirical terrain of this book. When amplification, personalisation, persistence, automation, and simulation operate under engagement incentives, token economies, or data-extractive business models, ambiguity hardens into recurring ethical tensions. This is the point at which affordance ambiguity becomes visible as the mechanism behind the dilemmas introduced in Chapter 3.

Progress versus preservation emerges wherever personalisation and simulation reconfigure identity, relationships, and social expectations. Social media expands connection but can erode slower norms of deliberation, forgiveness, and contextual charity. Web3 expands experimentation in coordination but can destabilise inherited safeguards embedded in legal and financial institutions. AI companions expand access to responsive interaction but risk reshaping expectations of reciprocity and care. In each case, progress alleviates some forms of suffering while unsettling practices that preservation seeks to protect. The dilemma persists because both poles represent legitimate commitments: to expand possibility, and to protect dignity.

Intention versus outcome becomes structurally salient because amplification and automation decouple purpose from effect. On social media, routine expression can be magnified into collective harm through engagement-optimised redistribution. In Web3, automated reward and governance rules can yield outcomes that no participant individually intended, yet that follow predictably from stake-weighted incentives. In AI companions, designers may aim to provide support, yet emergent conversational patterns can reinforce harmful trajectories when systems lack moral discernment. The dilemma is not reducible to "bad actors". It arises because the scale and automation of digital technologies systematically exceed individual foresight and local control.

Past versus future is embedded in persistence and path dependency. Digital traces bind present actors to prior decisions and prior selves. Ledgers and archives make early design choices durable. In Web3, early token allocations and governance rules imprint future power dynamics.[3] In social media, recommender architectures and data infrastructures create durable lock-in. In AI companions, accumulated conversational memory can become a durable asset that shapes future vulnerability and bargaining power. Ethical reflection must therefore operate in a time horizon longer than product cycles, confronting inherited constraints while anticipating long-term consequences under uncertainty.

Autonomy versus coordination emerges because personalisation expands agency while simultaneously generating interdependence that demands shared governance. Social media promises expressive freedom, yet collective safety requires enforceable norms and democratically controlled moderation. Web3 promises autonomy through permissionless participation, yet durable governance requires coordination, legitimacy, and mechanisms for dispute resolution. AI companions promise private, tailored interaction, yet systemic risk mitigation requires coordinated safeguards, auditing, and accountability structures. The structural tension is not between freedom and control as moral opposites, but between forms of autonomy that become socially costly and forms of coordination that risk becoming coercive or captured.

Unity versus diversity surfaces wherever amplification and personalisation reorganise publics. Social media simultaneously creates affinity communities and escalatory counterpublic conflict. Web3 ecosystems fragment into ideological sub-communities while competing for shared legitimacy and shared liquidity. AI companions can tailor worldviews and emotional scripts, potentially reinforcing psychological or ideological enclosure. The dilemma is not between connection and difference in the abstract, but between forms of unity that suppress pluralism and forms of diversity that erode shared ground.

The key lesson is that ethical dilemmas are the moral shape emerging when affordance ambiguity scales. When ambiguous action possibilities are actualised at planetary scale within commercial, tokenised, or data-extractive contexts, their dual potentials stabilise into recurring tensions. The dilemmas are not add-ons to technology adoption, nor are they unfortunate deviations from an otherwise benevolent trajectory. They are predictable, recurring outcomes of sociotechnical architectures that amplify, personalise, persist, automate, and simulate under asymmetrical power.

This bridge from affordance ambiguity to ethical dilemma prepares the ground for the remainder of this book. Chapter 3 will articulate the enduring tension types. The conceptual move here is to recognise that these dilemmas are not fully resolvable through optimisation or compliance, because they are produced by structurally ambiguous affordances operating within institutions that reward particular forms of actualisation. Chapters 5 through 7 will provide the empirical manifestations.

This is also why compassion is not introduced later as a sentimental afterthought. It is proposed in Chapter 4 as an ethical orientation suited to ambiguity: an orientation that keeps vulnerability and burden distribution in view when systems are designed to externalise harm, diffuse responsibility, and monetise attention or intimacy. Compassion does not dissolve dilemmas. It guides how we navigate them: where to introduce friction, what to preserve, what to redesign, and whose suffering should not be normalised as the acceptable price of innovation.

Why Compassion Is Not a Sentiment but a Governance Orientation

If affordances are inherently ambiguous and generate recurrent ethical dilemmas, then no compliance checklist can resolve them. Governance frameworks, safety audits, content policies, and technical safeguards are necessary, but they operate downstream of deeper sociotechnical logics. Amplification, personalisation, persistence, automation, and simulation are not accidental features that can simply be switched off or 'fixed'. They are constitutive of digital innovation. A platform may refine its moderation standards, a blockchain protocol may recalibrate token incentives, and an AI companion provider may publish updated safety guidelines. None of these measures abolish ambiguity. They merely reposition it.

Rule-following is thus not sufficient. Compliance can reduce liability and signal responsibility. It can align organisations with regulatory expectations. What it cannot do is substitute for a moral orientation capable of perceiving vulnerability where formal systems see only metrics. The deeper difficulty lies in the structure of the dilemmas themselves. Intention versus outcome cannot be eliminated because scale exceeds foresight. Once amplification and automation are embedded in infrastructures that serve millions or billions of users, consequences will outrun intention. Autonomy versus coordination cannot be resolved because meaningful agency presupposes collective safety, and collective safety constrains individual freedom. Progress versus preservation cannot be harmonised because innovation and stability express competing yet legitimate goods. In each case, decision-makers confront tensions that cannot be optimised away. Governance under such conditions requires more than procedural adherence. It requires a disciplined way of seeing and foreseeing.

This is where compassion becomes foundational, not as sentimentality, but as a constitutional orientation. In the philosophical framing developed in Chapter 4, compassion begins with sustained attention to suffering and is anchored in two imperatives: to hurt nobody and to help everybody as much as one can. These function as moral commitments that reshape how affordances are perceived and actualised. A purely instrumental orientation asks whether a feature increases engagement, efficiency, growth, or market share. A compassionate orientation asks different questions. Who bears the burdens generated by this design choice? Whose vulnerabilities are amplified? Whose exposure is rendered invisible? What forms of suffering become normalised as acceptable externalities of scale? Compassion interrupts the default mode of abstraction, optimisation, and moral distance through which previous waves of digital innovation unfolded. It insists that behind every data point lies a person whose dignity is not reducible to a performance metric.

In social media contexts, this shift reframes what counts as success. Instead of treating virality as an unqualified good, a compassionate orientation foregrounds the lived consequences of runaway exposure. It highlights the asymmetry between those who decide what gets amplified and those who are amplified. It asks whether frictionless sharing disproportionately burdens individuals with fewer resources to absorb reputational harm. It scrutinises moderation not merely for efficiency, but for fairness in how sanctions are distributed across differently situated users.

In Web3 infrastructures, compassion redirects attention from formal decentralisation to the distribution of risk and power. It questions whether permissionless participation genuinely includes those with minimal stake, or whether governance mechanisms subtly privilege early entrants and capital-rich actors. It asks whether transparency serves accountability for the many, or surveillance for the few. A compassionate lens does not reject programmable coordination. It interrogates whether rule-based automation helps the vulnerable or merely codifies existing power asymmetries.

In AI companions, compassion forces recognition that simulated empathy can coexist with structural indifference. Systems may be designed to respond warmly, remember preferences, and provide comfort. Yet if their underlying incentives reward prolonged engagement and data accumulation, the relational exposure created by confideability becomes morally dubious. Compassion insists that designers and regulators attend not only to the quality of responses, but to the risks embedded in persistent memory, automated counsel, and emotionally persuasive simulation. It foregrounds those least equipped to navigate these systems critically: adolescents, the lonely, and the distressed.

Compassion in this account is not reducible to empathy.[7] Empathy can be selective. It may concentrate moral attention on those who resemble us or whose suffering is emotionally vivid. Compassion, by contrast, requires sustained commitment to those structurally marginalised by design choices, incentive schemes, and infrastructural dependencies. It resists the comforting narrative that harm is exceptional or attributable only to deviant actors. When ambiguity is expressed at scale, predictable patterns of suffering follow. A constitutional orientation grounded in compassion recognises these patterns and treats their mitigation as a first principle rather than as damage control.

The ambition here is not to eliminate ambiguity. That would be neither feasible nor desirable. Digital affordances derive much of their value from openness, flexibility, and multiplicity. The task is to govern ambiguity responsibly. This entails introducing friction where amplification becomes destructive. It entails transparency mechanisms that render persistence accountable rather than merely durable. It requires auditing frameworks that interrogate automated decision-making not only for accuracy, but for distributive impact. It demands institutional arrangements that counterbalance simulation with safeguards capable of protecting those drawn into intimate disclosure.

It also requires vigilance against what might be called compassion-washing: the rhetorical performance of care without corresponding structural change. Platforms can deploy language of community and safety while maintaining engagement-maximising architectures. Web3 projects can invoke decentralised virtue while encoding stake-weighted control. AI companion providers can emphasise support while harvesting sensitive data. A compassionate governance orientation subjects such claims to scrutiny by examining whether burdens are genuinely shifted from the vulnerable to the powerful.

Compassion thus requires a willingness to embrace tensions constructively. Every design decision redistributes risk. Every optimisation privileges certain values over others. A compassionate stance does not pretend these redistributions can be eliminated. It demands that they be made visible and justified, with particular attention to those least able to bear additional harm. In this sense, compassion functions as a meta-ethical anchor within pluralistic societies. It does not prescribe a single ideological blueprint. It provides a stable reference point: avoidable suffering should not be normalised, dignity should not be sacrificed for efficiency, and gains should not disproportionately accrue to a privileged few.

The chapters that follow operationalise this claim. Chapter 3 will articulate the five recurring dilemma-types that arise from affordance ambiguity. Chapter 4 will deepen the account of compassion as a relational ethic capable of reorienting perception and action within these tensions. Chapters 5 through 7 will demonstrate how the dilemmas materialise in social media, Web3, and AI companions. The final part of this book will explore how individuals, organisations, and public institutions can embed compassionate principles into everyday practice and infrastructural design.

This chapter has established the mechanism. Digital technologies matter ethically not because they occasionally malfunction, but because their affordances are inherently ambiguous and are actualised at scale within institutional environments structured by concentrated power and misaligned incentives. Ambiguity crystallises into dilemmas. Dilemmas cannot be optimised away, only navigated in more or less ethical ways. Compassion is proposed as a foundational orientation, not as soft sentiment, but as disciplined attention to suffering in systems designed to obscure it.

References

1 Ciriello, R. F., Gal, U., Hannon, O. & Thatcher, J. Responsible social media use: how user characteristics shape the actualisation of ambiguous affordances. *European Journal of Information Systems* **34**, 799–821, doi:10.1080/0960085X.2024.2444249 (2025).

2 Ciriello, R. F. Dear Schopenhauer: how can IS artefacts help us flourish? *Wille, Vorstellung,* compassion, and beauty in social media. *Scandinavian Journal of Information Systems* **37**, 317–354, DOI:10.2139/ssrn.5285180 (2025).

3 Viguerie, C., Ciriello, R. F., Zavolokina, L. & Mathiassen, L. Explaining decentralisation in enterprise blockchain governance: the imprint of platform openness and

participant inclusiveness. *Journal of Information Technology*, 02683962261422482, doi:10.1177/02683962261422482 (2026).

4 Ciriello, R. F., Chen, A. Y. & Rubinsztein, Z. A. Compassionate AI design, governance, and use. *IEEE Transactions on Technology and Society* **6**, 1–6, doi:10.1109/TTS.2025.3538125 (2025).

5 Ciriello, R. F., Marx, J., Cheong, M., Mueller-Bloch, C. & Mathiassen, L. Decentralized social media. *Business & Information Systems Engineering* **68**, 219–234, doi:10.1007/s12599-025-00952-4 (2025).

6 Klincewicz, M., Alfano, M. & Fard, A. E. Slopaganda: the interaction between propaganda and generative AI. *arXiv preprint arXiv:2503.01560* (2025).

7 Bloom, P. *Against Empathy: The Case for Rational Compassion.* (Random House, 2017).

3 Ethical Dilemmas of Sociotechnical Change

Can you imagine a life without the internet? A growing number of young people now wish for one. They romanticise the pre-internet age, turn to "dumb phones", experiment with digital detoxes, organise "digital luddite" meetups, and speak openly about wishing social media had never been invented.[1] Their nostalgia is not simply sentimental or escapist. It reflects a widespread sense that digital technologies have reshaped everyday life faster than individuals can adapt. Many are tired of being told that technology equals progress, especially technologies they neither asked for nor shaped.

At the same time, societies find themselves unable to step away. Governments around the world are trying to rein in the power of corporate platforms. The European Union debates digital sovereignty, seeking meaningful public alternatives to privately owned platforms. Australia's world-first 2025 Social Media Minimum Age law, which obliges platforms to take reasonable steps to prevent under-16s from holding accounts, has triggered expert praise, public protest, and constitutional challenge.

Similar oscillations between dependence and resistance appear across domains. Few people can participate in modern societies without relying on the infrastructures of Meta, Google, Microsoft, or Amazon. Even academics who critique digital power do so through search engines, evaluation systems, and journal infrastructures controlled by a handful of global publishers. This book is published by one of them, and is therefore part of the very machine it critiques.

This cultural longing for disconnection, paired with deep structural dependence, reveals a deep-seated contradiction in sociotechnical change. Digital technologies become entwined with the conditions of social, economic, and political life even as they generate discomfort, fatigue, and harm. We want the benefits they provide, yet resent the harms they create. We seek protection from their excesses, yet rely on them for communication, coordination, and recognition. It often feels as though we can neither live with these technologies nor without them.

That tension between desire and disillusionment, between reliance and resistance, is not a superficial contradiction but the hallmark of an ethical dilemma.

DOI: 10.4324/9781003614814-4

Ethical Dilemmas: More Than Technical Optimisation Problems

Ethical dilemmas arise when legitimate moral obligations collide in ways that make it impossible to satisfy each to the same extent. Neither obligation is inherently superior. Instead, communities and institutions confront situations in which every available course of action fulfils one set of commitments while compromising another. Digital innovation is especially prone to such tensions. Sociotechnical change reconfigures how people communicate, work, learn, and relate at a pace that outstrips the ability of social norms and institutions to adapt. It amplifies inequalities, introduces new dependencies, and unevenly distributes benefits and burdens. Ethical dilemmas therefore rarely appear as simple "right or wrong" choices, but as collisions between defensible values whose significance varies with status, resources, and power.

These conflicts do not arise from features alone, but from ambiguous affordances that are actualised at scale within institutional contexts (Chapter 2). Under conditions of scale, opacity, and concentrated governance (Chapter 1), value trade-offs become durable rather than episodic. Ethical dilemmas are therefore not accidental by-products of digital innovation, nor are they amenable to technical optimisation. They are constitutive features of sociotechnical systems: interdependent networks of people, technologies, institutions, and cultural expectations at scales no single actor controls. Decisions made by designers, firms, policymakers, and users interact in ways that cannot be traced to any singular intention or act. When billions participate in digital infrastructures, no vantage point can fully represent the range of values at stake. What appears as progress to some may feel like dispossession to others; what looks efficient from one angle may appear as surveillance or precarity from another.

Technical optimisation alone is bound to fail because it assumes stable goals and commensurable outcomes, whereas ambiguous affordances generate incommensurable goods and harms that shift with context and power. Ethical dilemmas thus demand pluralistic interpretation and ongoing negotiation rather than a definitive solution. Pluralism here does not mean moral permissiveness. It means recognising multiple legitimate claims while still treating dignity and avoidable suffering as non-negotiable constraints.

Part of the reason ethical dilemmas persist is that sociotechnical change is inherently value-reconfiguring. Digital technologies reshape visibility, agency, memory, labour, intimacy, and knowledge. In doing so, they also reshape the moral expectations attached to each domain. They alter who can see whom, who can act when and how, what is recorded and for how long, how people present themselves, and how decisions are justified. Such changes do not merely accentuate established values. They modify their meaning and relative priority. When persistence makes past speech permanently retrievable, privacy becomes less about secrecy than about how contexts collapse. When personalisation predicts what we want, autonomy becomes entangled with engineered attention. Trust, likewise, becomes mediated through interfaces rather than institutions. Ethical dilemmas therefore emerge because people must navigate value landscapes that are shifting, contested, and only ever partially compatible with inherited norms.

Unpredictability further complicates ethical judgement. Sociotechnical systems produce emergent and non-linear effects that often diverge from intended designs. Small interactions can cascade into large consequences. Modest interventions can produce disproportionate change. These dynamics intensify at the global scale of digital platforms, where billions of daily interactions traverse linguistic, cultural, economic, and political boundaries.[2] Benefits and harms alike are magnified, and ethical consequences often exceed designers' foresight. Policymakers can anticipate some outcomes, but often cannot reliably predict how interventions will interact with existing power structures or cultural practices.

Attempts to eliminate dilemmas through optimisation, nudging, or automation typically generate new tensions. Efforts to reduce misinformation may compromise freedom of expression. Interventions that strengthen safety may constrain autonomy. Mechanisms that enhance coordination may marginalise minority perspectives. Pursuing any single value to perfection destabilises another. Australia's Robodebt scheme, designed to streamline welfare compliance through algorithmic debt calculation, promised efficiency yet produced devastating harms by replacing human discretion with automated certainty.[3] The proposed automation of the National Disability Insurance Scheme's budget-setting risks repeating the same logic: privileging throughput over contextual judgement, narrowing avenues for human review, and placing the greatest burdens on those with complex or poorly captured needs.[4] The point is not simply that automation can backfire, but that it intensifies ethical dilemmas by replacing contextual judgement with scalable rule execution.

None of this means that ethical action is futile. It means dilemmas arise because sociotechnical systems distribute capacities and vulnerabilities unevenly across groups whose interests and experiences cannot be reconciled through any single logic.

Understanding ethical dilemmas as constitutive of sociotechnical change shifts the task from seeking definitive solutions to cultivating pluralistic ethical judgement. The dilemmas discussed in this chapter reveal the moral complexity of sociotechnical systems. They bring into view the values at stake, the power relations that shape their pursuit, and the vulnerabilities exposed when systems cause harm. They also highlight the limits of technocratic fixes. Clever engineering can mitigate specific harms but cannot adjudicate competing moral claims. Good regulation can set boundaries but cannot eliminate the tensions inherent in infrastructures serving diverse constituencies.

This chapter therefore treats dilemmas as analytical lenses rather than problems to solve. They help us examine how value conflicts are produced, how responsibilities are distributed, and how sociotechnical arrangements shape the possibilities for ethical action. The five dilemmas introduced next show why these conflicts remain unsettled and often irreconcilable, and they prepare the ground for understanding why compassion becomes essential for navigating them in ways that respect human dignity and acknowledge the uneven burdens of sociotechnical change.

A Dialectical View on Ethical Dilemmas

Dialectical thinking, commonly associated with Hegel and Marx but with older roots in Greek philosophy, begins from a simple insight: many forces shaping human experience pull in opposite directions yet remain interdependent.[5] Dialectics offers a vocabulary for understanding how actors respond when ambiguous affordances institutionalise enduring value tensions. Progress and preservation, autonomy and coordination, and unity and diversity are not discrete choices but mutually constitutive dynamics that shape and reshape one another over time.

Dialectics therefore treats oppositions not as errors to be corrected but as engines of sociotechnical change that reveal what societies value, who benefits from innovation, and who bears its burdens. This perspective departs from optimisation metaphors that assume a stable equilibrium can be reached. In practice, every attempt to privilege one value – openness, privacy, control, frictionlessness – reconfigures the landscape and produces new tensions. Digital technologies intensify these dynamics by reorganising visibility, speed, agency, and scale. They make contradictions harder to ignore and ensure that even well-intentioned reforms trigger second-order effects that reverberate widely.

A dialectical view also highlights patterned responses to dilemmas. Four recur: suppression, suspension, separation, and synthesis. Suppression ignores a dilemma, often strengthening it by pushing it out of sight until it resurfaces with greater force. Suspension defers engagement in the hope that ambiguity will resolve itself, a familiar reaction to fast-moving technologies whose implications remain uncertain. Separation isolates opposing values – for instance, assigning privacy to compliance teams and innovation to product groups – yet this typically entrenches either/or thinking and fosters conflict. Synthesis, the most demanding response, treats the opposition itself as a resource for reimagining practice: integrating human discretion into automated systems; combining decentralised architectures with shared safety standards; and designing platforms that preserve cultural expression while enabling responsible innovation. None of these responses resolves a dilemma. Rather, they shape how it evolves and whose interests prevail. Some produce vicious cycles in which narrow framings ossify into dogma; others create expansive possibilities by engaging opposing values on their own terms. Digital innovation offers abundant examples of both.[5]

These dynamics clarify why ethical dilemmas are never merely technical optimisation problems. They surface deeper struggles over legitimacy, accountability, and control. They reveal how power operates through design choices, default options, data practices, and institutional arrangements, often invisibly to those who are affected by them. Vulnerability sits at the centre of this picture. It is not a personal trait but a relational condition produced by asymmetries in information, resources, and influence. Engagement-optimised feeds, addictive design patterns, and opaque algorithms do not distribute risk evenly, but concentrate it among those least able to resist or exit.

As digital technologies become entrenched, early design compromises harden into structural constraints. Practices once contested can normalise with surprising speed. Public acceptance of remote drone warfare, once heavily debated but now embedded in geopolitical routines, illustrates how quickly alternatives recede once systems scale. With AI still in flux, societies face a narrowing window to shape its ethical trajectory deliberately rather than retrospectively.

Dialectics prepares the ground for compassion by directing ethical reflection toward those most exposed to harm and by reframing dilemmas not as zero-sum contests but as shared moral and political struggles. It invites readers to approach the five dilemmas that follow not as puzzles awaiting resolution but as enduring sites of negotiation where more humane futures remain possible. These tensions point toward the need for an ethical orientation attuned to suffering and vulnerability and capable of engaging contradictions without reducing them. Compassion, developed in the next chapter, offers precisely such an orientation.

Five Foundational Ethical Dilemmas of Sociotechnical Change

The dilemmas introduced in this section form the intellectual centre of this book. They recur across domains because they express tensions that cannot be resolved once and for all. Each technological wave reconfigures them in new and often unexpected forms. Table 3.1 provides an overview.

Dilemma 1: Progress versus Preservation

The tension between progress and preservation is among the oldest ethical conflicts. Progress matters because the world is marked by structural inequities, fragile institutions, unsustainable practices, and avoidable suffering. Innovation can remedy such conditions. Although often framed in economic terms, the deeper imperative for progress is moral rather than commercial: a commitment to improve collective wellbeing and reduce unjustified harm. This imperative underwrites advances in sustainability, public health, accessibility, education, and democratic participation.

Table 3.1 Five Foundational Ethical Dilemmas of Sociotechnical Change

Dilemma	*Definition*	*Examples*
Progress vs Preservation	*Progress* pursues innovation to enhance wellbeing, efficiency, access, or justice. *Preservation* safeguards traditions, rights, identities, and institutional or ecological stability. The tension arises because change that reduces suffering can also erode continuity or hard-won freedoms.	**Social media:** Expanded participation and voice, yet destabilised deliberative norms and externalised harms such as polarisation and mental health decline. **Web3:** Enables novel coordination and ownership, yet disrupts legal safeguards and fuels speculative excess. **AI companions:** Expand emotional accessibility and creativity, yet commodify intimacy and unsettle relational traditions.
Intention vs Outcome	*Intention* concerns the purposes of designers, firms, or users. *Outcome* concerns systemic effects that emerge at scale. The tension arises because sociotechnical systems decouple purpose from consequence, rendering harms foreseeable yet difficult to attribute.	**Social media:** Built to connect, yet amplified misinformation and harassment through engagement-optimised architectures. **Web3:** Designed for decentralisation, yet produces stake-weighted concentration and elite capture. **AI companions:** Created for support, yet can reinforce dependency or generate harmful advice when automation exceeds discernment.
Past vs Present vs Future	The *past* comprises inherited institutional and technical imprints. The *present* is action under constraint. The *future* reflects path-dependent yet uncertain consequences. The tension arises because current decisions reproduce historical structures while shaping trajectories under partial predictability.	**Social media:** Early ownership and design choices continue to shape moderation and power asymmetries. **Web3:** Initial token allocations and consensus mechanisms lock in governance dynamics. **AI companions:** Today's data retention and emotional modelling decisions will shape future norms of intimacy and oversight.
Autonomy vs Coordination	*Autonomy* denotes self-determination and freedom of action. *Coordination* denotes shared rules and institutions enabling stability. The tension arises because	**Social media:** Expressive freedom clashes with coordinated moderation to prevent abuse and misinformation. **Web3:** Permissionless participation fosters experimentation,

(*Continued*)

Table 3.1 (Continued)

Dilemma	*Definition*	*Examples*
	agency depends on coordinated infrastructures, yet coordination constrains freedom and risks capture.	yet sustainability requires interoperable standards and dispute resolution. **AI companions:** Personalised interaction enhances agency, yet safety demands coordinated safeguards and regulation.
Unity vs Diversity	*Unity* provides shared norms and common ground; *diversity* sustains plural perspectives and moral difference. The tension arises because coherence can suppress difference, while unmediated pluralism fragments publics and weakens the commons.	**Social media:** Connects diverse communities, yet algorithmic amplification of moral conflict intensifies polarisation. **Web3:** Enables plural sub-communities, yet struggles to maintain shared legitimacy. **AI companions:** Hyper-personalised scripts affirm identity, yet risk insulating users within psychologically or ideologically closed worlds.

Preservation matters no less. Communities organise their moral worlds through traditions, languages, customs, and cultural artefacts that confer identity and meaning. These are not nostalgic residues but sources of continuity, resilience, and situated knowledge that often guard against domination and unforeseen harm. Conservationists defending ecosystems from extractive development and Indigenous communities protecting sacred sites from destructive infrastructure illustrate that preservation is a moral commitment rather than sentimental attachment. History repeatedly shows that change, even when well-intentioned, can generate harms visible only in retrospect.

Neither pole is unambiguously virtuous. Innovation can erode cultural memory or weaken social bonds. Preservation can entrench exclusion or impede the flourishing of those long denied recognition. The dilemma arises because both reflect legitimate commitments that frequently collide.

Digital innovation sharpens this tension by reconfiguring entire spheres of life with unprecedented speed and asymmetry. Social media illustrates the pattern. What began as playful tools for connection evolved into infrastructures underpinning communication, politics, and commerce. While these platforms became central to democratic life, benefits were largely privatised and burdens – misinformation, harassment, polarisation, mental health decline – externalised onto society.[6]

Generative AI intensifies the dilemma. It enables creativity, accessibility, and educational support, yet destabilises rights that preservation seeks to protect. Non-consensual deepfake pornography, disproportionately targeting women, shows how advances in synthetic media collide with dignity and safety.[7] Disputes over synthetic voices and copyright litigation reveal how creative labour can be appropriated at scale.[8] Training data skewed toward Western industrialised contexts marginalises epistemic traditions and languages of the global South, risking gradual cultural erosion.[9]

This dilemma animates later chapters across social media, Web3, and AI companions. Each domain promises new forms of coordination or care while threatening values that communities seek to sustain. The task is not to choose progress over preservation, but to recognise that digital innovation continually reshapes their boundaries. Ethical judgement therefore requires attention to what is gained, what is lost, and who bears the cost.

Dilemma 2: Intention versus Outcome

The second dilemma concerns the gap between intention and outcome. The familiar saying that the road to hell is paved with good intentions captures a general truth about social life: moral purpose does not guarantee moral effect. Economic reforms pursued in the name of productivity can deepen social exclusion or accelerate environmental degradation. Medical interventions designed to relieve suffering can impose profound burdens on caregivers or animals used for drug testing. A person may act with malice and yet produce a benign outcome, just as someone acting with care may still cause serious harm.

Ethical judgement therefore cannot rely on intentions alone. Even in legal domains, where responsibility is tied to voluntary choice and foreseeability, boundaries blur in practice. Digital contexts magnify this difficulty because sociotechnical systems disperse agency, obscure causal chains, and accelerate feedback loops.

Digital technologies magnify this difficulty because sociotechnical systems disperse agency, obscure causal chains, and accelerate feedback loops. The Therac-25 radiation accidents remain instructive. A constellation of software flaws, interface ambiguities, and organisational misunderstandings led to patient deaths, though no single developer intended harm.[10] Individually reasonable actions converged into catastrophic system-level failure. Under opacity, responsibility diffuses. Under concentration, the power to set defaults and incentives remains asymmetrical. "Unintended consequences" thus become predictable governance patterns rather than surprising lapses.[11]

Contemporary digital infrastructures exhibit similar dynamics. Social media amplifies speech through individually trivial but collectively potent interactions. A playful video can metastasise into a harassment campaign within hours. Viral harms arise from the interplay of user behaviour, algorithmic curation, platform incentives, and cultural atmospheres rather than any isolated act.[2]

The language of unintended consequences is often too thin.[1] Outcomes may be unintended, yet not unforeseeable. Bias in hiring algorithms, discriminatory facial recognition, and the targeting of women in deepfake pornography have been repeatedly documented.[12] Their persistence reflects not only complexity but also tolerance of predictable harm. Early social media rhetoric promised openness and connection; outcomes have included enclosure, surveillance advertising, loneliness, and polarisation. These effects are not mere accidents. They reflect design choices shaped by commercial incentives, path dependencies, and entrenched asymmetries.

The dilemma rests on the tension between moral purpose and emergent effects. Digital innovation unfolds in environments where outcomes are path-dependent and relational, only loosely coupled to original intent. As later chapters show across social media, Web3 governance, and AI companionship, ethical responsibility requires attention not only to what individual actors mean to do but also to what systems actually do when components interact at scale.[13]

Dilemma 3: Past versus Present versus Future

The third dilemma (or trilemma) concerns how we orient ourselves across time. A well-known adage attributed to Gautama Buddha reminds us that "*The past is already gone, the future is not yet here. There's only one moment for you to live, and that is the present moment*". People act in the present, yet they are shaped by inherited structures and assumptions about what the future should hold. Sociologists refer to this as imprinting: early structures become sedimented, guiding decisions long after the original context has faded.[14] Religious orders that persist for centuries, constitutional arrangements still expressing the priorities of their founders, and bureaucratic routines that outlive their purpose all illustrate how the past endures. Digital infrastructures exhibit similar persistence. Dual-class share structures at Meta and Alphabet entrench the influence of their founders; early platform and protocol design choices generate path dependencies that are difficult to reverse.[15]

Acting in the present thus means navigating inherited constraints while attempting to create futures that do not simply reproduce them. Current debates over artificial intelligence reflect this tension. Incentives for rapid deployment coexist uneasily with historical lessons about regulatory failure and uncertain long-term risk. Web3 ecosystems display comparable dynamics, where early consensus mechanisms or token allocations shape what governance can plausibly become. Constitutional analogies clarify the pattern: eighteenth-century compromises in the United States still shape political representation today.[16] Digital infrastructures replicate this dynamic at far greater speed and scale.

The future remains uncertain, yet not wholly unpredictable. Complex systems such as weather patterns or financial markets can be forecast in broad strokes despite short-term volatility. Digital innovation displays a similar structure. It is easier to anticipate macro-trajectories – consolidation of platform power, expansion of algorithmic mediation – than the specific consequences of particular tools. This mixture of partial predictability and deep uncertainty complicates ethical judgement. It demands attentiveness to historical lessons, awareness of inherited structures, and humility about what cannot be foreseen.

Dilemma 4: Autonomy versus Coordination

The fourth dilemma concerns the tension between autonomy and coordination. Autonomy reflects a basic human need for agency, self-determination, and the capacity to shape one's environment. Across cultures and eras, the impulse toward freedom has been among the most powerful forces in human life. People resist domination, rebel against imposed constraints, and at times fight for the ability to direct their own trajectories. This drive underpins democratic societies, civil rights movements, and everyday refusals to be told how to live. Autonomy is therefore not a luxury but a condition of dignity and human flourishing.[17]

Coordination is equally fundamental. Humans dominate the planet not because we are strongest or fastest (or even smartest), but because we coordinate through shared symbols, narratives, and institutions. Language, norms, laws, and cultural practices enable collective action at scale.[18] These forces are not opposites but interdependent: autonomy fosters experimentation and diversity, while coordination sustains stability and shared purpose.

Digital innovation places these forces in delicate relation. Individuals can act with unprecedented freedom, yet their actions depend on infrastructures that require extensive coordination. Web3 ecosystems illustrate this vividly. Their ethos prizes autonomy through permissionless entry, open-source code, and the right to dissent. Yet these same features can generate fragmentation and make institutional coherence difficult to sustain.

The history of rail transport offers a useful analogy. Nineteenth-century societies operated with incompatible track gauges, producing inefficiencies that persisted until standardisation became unavoidable. Decentralised digital infrastructures face similar pressures. Consensus mechanisms, identity systems, and security architectures must interoperate to ensure safety and usability, yet imposing standards risks constraining the autonomy that attracts participants in the first place.

The dilemma extends beyond decentralised systems. Social media platforms rely on coordinated norms to limit harassment and misinformation, yet rigid enforcement can suppress legitimate expression and cultural plurality.[19] AI development requires shared safety practices to mitigate systemic risk, yet firms compete to preserve autonomy in product design and release.

Democratic governance mirrors this tension. Citizens value freedom of expression and association, yet stable collective decisions require coordination and constraint. Pushing either pole to its extreme produces pathologies. Excessive autonomy fosters fragmentation, exploitation, and capture by well-resourced actors. Excessive coordination suppresses plurality, entrenches hierarchy, and weakens the creative and deliberative capacities on which innovation depends.

Because neither pole can be eliminated, societies must continually renegotiate their balance. Later chapters show how these tensions manifest in social media governance, Web3 architectures, and AI companionship. They also clarify why compassion becomes relevant. Compassion foregrounds vulnerability and interdependence. It reminds us that autonomy is meaningful only when it does not harm others, and that coordination is legitimate only when it respects the agency of those it affects. It encourages governance that treats autonomy and coordination not as absolutes but as values to be held in productive tension if digital futures are to remain humane.

Dilemma 5: Unity versus Diversity

The fifth dilemma concerns the tension between unity and diversity. Both are widely valued and essential for democratic societies, yet they often conflict. National metaphors capture this contrast. The United States has long imagined itself as a melting pot, welcoming newcomers while expecting adherence to a shared moral code. Canada and Australia lean more toward a mosaic ideal, where plural perspectives coexist without dissolving into uniformity. Each model reflects a stance on belonging and difference. Unity offers coherence, mutual recognition, and the capacity for collective action. Diversity fosters creativity, adaptability, and the inclusion of voices that might otherwise be marginalised.

Digital innovation intensifies these tensions by bringing disparate groups into sustained and often unmediated contact. It exposes individuals to ways of living that challenge inherited norms, creating opportunities for learning but also for conflict. Jonathan Haidt's work on moral psychology helps explain why such encounters can become emotionally charged and destabilising.[20] He distinguishes demographic diversity, often linked to positive outcomes, from moral diversity, which can provoke discomfort or ostracism. People differ in the moral foundations that shape their judgements about authority, loyalty, purity, fairness, and care. These differences can be so pronounced that individuals feel they inhabit distinct moral universes when interpreting the same event. This account is best read as a heuristic rather than a comprehensive explanation.

A classic study from the 1960s illustrates the point. Under Jim Crow segregation, white Christians in the southern United States reported greater social comfort with Black Christians than with white atheists.[21] Shared belief mattered more than shared race. Moral alignment outweighed demographic similarity. Digital technologies scale such dynamics dramatically. Social media exposes individuals to vast moral diversity with little contextual scaffolding, generating misunderstanding, outrage, and cycles of polarisation.[2] Online spaces often become arenas where moral boundaries are tested, contested, and sometimes weaponised.

The ethical question is not whether unity or diversity is preferable. Both are necessary for democratic and pluralistic life. The challenge is to sustain shared ground without erasing difference, and to uphold pluralism without fragmenting the commons. Certain values – fairness, reciprocity, justice, compassion – appear to enjoy broad cross-cultural resonance and underpin near-universal prohibitions against murder, rape, and fraud. Other practices, such as drinking, gambling, or same-sex relationships, vary across time and context, often independently of legal change.[13] Digital platforms collapse these contextual boundaries, placing diverse moral systems into direct interaction. This can enrich public life, but also destabilise it, making unity feel coercive and diversity feel threatening. The unity–diversity dilemma thus concerns how to build common ground without demanding uniformity, and how to preserve pluralism without sliding into hostility or fragmentation.

Navigating Ethical Dilemmas, Compassionately

The five dilemmas outlined in this chapter are not an exhaustive taxonomy. They serve as a heuristic for framing recurrent problems in the ethical landscape of sociotechnical change. Digital technologies generate additional tensions around labour, environmental sustainability, and the political economy of data, among others. Yet the dilemmas identified here are especially salient because they centre on competing goods that cannot be fully satisfied at once. They recur across domains, appearing in distinct forms in platform governance, algorithmic design, Web3 coordination, and the intimate entanglements of AI companions. They help explain why digital innovation generates persistent controversies that stem not merely from poor design, bad use, or weak regulation, but from deeper value conflicts that cannot be eliminated, only navigated.

What these dilemmas share is that they arise where power asymmetries shape both universal human vulnerability and its unequal distribution. They expose how digital infrastructures institutionalise decision rights, redistribute risks, and define whose values "count". Progress and preservation conflict not because innovation is inherently destabilising, but because groups differ in their influence over what is preserved and what fades. Intention and outcome diverge most sharply when designers are insulated from the consequences of their systems. Past, present, and future collide when decision-making authority rests with actors whose interests may not align with those downstream. Autonomy and coordination strain when well-resourced actors can reshape governance outcomes. Unity and diversity clash when platforms privilege certain moral sensibilities while marginalising others. These are not abstract puzzles but concrete expressions of structural inequality, contested legitimacy, and uneven voice.

It is tempting to treat such dilemmas as optimisation problems, inviting the search for equilibrium between competing values. Yet ethical dilemmas are inherently unresolvable. They are not temporary frictions en route to stability, but structural features of complex systems. Attempts to resolve them through technical fixes, local optimisation, or universal rules typically generate new tensions. Strengthening privacy may constrain personalisation. Coordinating global standards may marginalise local practices. Moderation rules designed to safeguard civil discourse may suppress legitimate dissent. These tensions persist because they reflect enduring disagreements about harm, justice, freedom, and responsibility. Ethical navigation therefore requires ongoing deliberation rather than a definitive solution.

A further thread running through these dilemmas is vulnerability. Digital innovations do not affect all individuals equally. Vulnerability is not a personal deficit, but a situational, relational, and systemic condition. It depends on access to resources, the ability to represent one's interests, and the capacity to contest harmful practices. It is also engineered. Engagement-optimised feeds disproportionately affect those with limited time, resilience, or social support. Opaque recommendation systems shape behaviour through signals users cannot meaningfully challenge. Those most affected by digital harms often have the least influence over design and governance. Children targeted by addictive interfaces, women subjected to deepfake abuse, gig workers managed by algorithmic scoring, and communities exposed to predictive policing rarely have a seat at the table where decisions about data, incentives, and rules are made. Ethical dilemmas thus fall hardest on those whose vulnerabilities are amplified by digital infrastructures.

Vulnerability is not only personal but institutional. Once embedded, digital infrastructures are difficult to unwind. Early decisions about protocols, algorithms, and governance create path dependencies that constrain future possibilities. A stark analogue lies outside the digital sphere. Two decades ago, the United States introduced targeted drone strikes under narrow national security authorisations. Within years, their use expanded and was reframed as a precise and humane instrument of war. Today, drone strikes are routine despite ongoing civilian casualties. The controversy has faded even as the ethical questions persist.[22] A similar drift characterises digital infrastructures. Social media platforms have evolved into quasi-public utilities shaping political discourse, youth wellbeing, and cultural production. Reversing these trajectories now requires confronting immense economic, technical, and social barriers.

Recognising path dependence is critical for emerging technologies such as artificial intelligence. The present moment remains a narrowing window of malleability. Norms, governance structures, and institutional expectations for AI are not yet fixed. There is still an opportunity to shape how these systems integrate into education, healthcare, public administration, and everyday intimacy. The dilemmas introduced here carry directly into AI design, where early decisions about data governance, emotional modelling, and companion-like behaviour will lock in assumptions about what forms of human flourishing are valued. AI need not replicate the trajectory of earlier platforms, but it will if ethical reflection remains at the level of surface trade-offs rather than engaging power and vulnerability more deeply. The same holds for future waves of digital innovation.

This is where compassion enters as a constitutional orientation. Compassion does not dissolve dilemmas, nor does it offer an algorithm for resolving competing values. Its contribution lies elsewhere. It foregrounds the lived experience of those most affected by sociotechnical change. It asks who is vulnerable, whose perspectives remain unheard, and which harms prevailing narratives obscure. It shifts ethical reasoning away from technocratic efficiency toward the recognition of suffering and the responsibility to alleviate it. In doing so, compassion reframes dilemmas not as zero-sum contests but as shared moral challenges. It invites proportional responsibility, encouraging designers, policymakers, and organisations to consider both their capacity to shape outcomes and their obligations toward those who bear the greatest risks.

In its pluralistic form, developed in the next chapter, compassion also supports more legitimate value reconfiguration. When progress and preservation collide, compassion directs attention to whose histories and identities are at stake. When intentions diverge from outcomes, it demands accountability for foreseeable harms. When past, present, and future come into tension, it offers a lens for imagining futures that do not reproduce historical injustice. When autonomy and coordination strain, it asks how to safeguard agency while protecting vulnerable groups. When unity and diversity clash, it asks how to sustain common ground without silencing minorities. Compassion thus operates not as sentiment but as an ethical anchor for deliberation in complex, contested environments.

This orientation prepares the ground for the next chapter, which develops compassion as a pluralistic ethical framework. Compassion does not replace rights, rules, or ethical reasoning. It complements them by centring vulnerability, relationality, and human dignity. It counters the ethos of optimisation, extraction, and individualisation that shapes much contemporary digital innovation. It urges a shift from narrow efficiency metrics toward broader questions of legitimacy, justice, and shared responsibility. As the next chapter shows, compassionate digital innovation offers a pathway toward value reconfigurations that are not only technically feasible but socially and morally sustainable. The discussion now turns to this broader ethical foundation and to Schopenhauer's account of vulnerability, striving, and representation, which illuminates why the ethical dilemmas of sociotechnical change arise in the first place.

References

1 Ciriello, R. F., Taylor, F. & Hassan, R. *Imagining a life without the internet: have you tried to step back from the screen?*, https://www.abc.net.au/listen/programs/lifematters/life-without-the-internet/105344118 (ABC Radio: National, 2025).

2 Ciriello, R. F., Gal, U., Hannon, O. & Thatcher, J. Responsible social media use: how user characteristics shape the actualisation of ambiguous affordances. *European Journal of Information Systems* **34**, 799–821, doi:10.1080/0960085X.2024.2444249 (2025).

3 Marjanovic, O., Cecez-Kecmanovic, D. & Vidgen, R. Theorising algorithmic justice. *European Journal of Information Systems* **31**, 269–287, doi:10.1080/0960085X.2021.1934130 (2022).

4 Van Toorn, G. *Did we learn nothing from Robodebt? NDIS automation will put vulnerable lives at the mercy of machines*, https://www.theguardian.com/commentisfree/2025/dec/04/ndis-news-support-plans-automated-robodebt-national-disability-insurance-scheme (2025). Accessed 17 February 2026.

5 Ciriello, R. F. & Mathiassen, L. in *International Conference on Information Systems* (Copenhagen, Denmark, 2022).

6 Mueller-Bloch, C. & Ciriello, R. Governing social media as a public utility: a case for sovereign digital infrastructure. *Communications of the ACM* forthcoming, https://doi.org/10.48550/arXiv.2602.12535 (2026).

7 Shirish, A. & Komal, S. A socio-legal inquiry on deepfakes. *California Western International Law Journal* **54**, 517 (2023).

8 Milmo, D. *Scarlett Johansson's OpenAI clash is just the start of legal wrangles over artificial intelligence*, https://www.theguardian.com/technology/article/2024/may/27/scarlett-johansson-openai-legal-artificial-intelligence-chatgpt (The Guardian, 2024). Accessed 17 February 2026.

9 Atari, M., Xue, M., Park, P., Blasi, D. & Henrich, J. Which humans?, doi:10.31234/osf.io/5b26t (2023).

10 Birsch, D. Moral responsibility for harm caused by computer system failures. *Ethics and Information Technology* **6**, 233–245, doi:10.1007/s10676-005-5609-5 (2004).

11 Ciriello, R. F., Mathiassen, L., Risius, M., Cheong, M., Scheepers, H., & Vaast, E. Compassionate Digital Innovation: A Pluralistic Perspective and Research Agenda. *Information Systems Journal* **36**, 364–385, doi:10.1111/isj.70003 (2025).

12 Ciriello, R. F. & Kögel, S. I. in *Australasian Conference on Information Systems (ACIS2024)* (Canberra, Australia, 2024).

13 Stahl, B. C. Responsible computers? A case for ascribing quasi-responsibility to computers independent of personhood or agency. *Ethics and Information Technology* **8**, 205–213, http://doi.org/10.1007/s10676-006-9112-4 (2006).

14 Stinchcombe, A. L. Social structure and organizations. In *Handbook of Organization* (ed. March, P) 142–193 (Rand McNally, 1965).

15 Viguerie, C., Ciriello, R. F., Zavolokina, L. & Mathiassen, L. Explaining decentralisation in enterprise blockchain governance: the imprint of platform openness and participant inclusiveness. *Journal of Information Technology*, 02683962261422482, doi:10.1177/02683962261422482 (2026).

16 Ciriello, R. F. & Mindel, V. Techno-collectivism: scaling libertarian socialist governance of digital infrastructure. *IEEE Technology and Society Magazine*, 1–15, doi:10.1109/MTS.2025.3605115 (2025).
17 Alexander, G. S. Property, dignity, and human flourishing. *Cornell Law Review* **104**, 991–1046 (2019).
18 Bingham, P. M. Human uniqueness: a general theory. *The Quarterly Review of Biology* **74**, 133–169, doi:10.1086/393069 (1999).
19 Ciriello, R. F., Marx, J., Cheong, M., Mueller-Bloch, C. & Mathiassen, L. Decentralized social media. *Business & Information Systems Engineering* **68**, 219–234, doi:10.1007/s12599-025-00952-4 (2025).
20 Haidt, J., Rosenberg, E. & Hom, H. Differentiating diversities: moral diversity is not like other kinds. *Journal of Applied Social Psychology* **33**, 1–36, doi:10.1111/j.1559-1816.2003.tb02071.x (2003).
21 Rokeach, M. & Mezei, L. Race and shared belief as factors in social choice. *Science* **151**, 167–172, doi:10.1126/science.151.3707.167 (1966).
22 Krebs, S. Above the law: drones, aerial vision and the law of armed conflict – a socio-technical approach. *International Review of the Red Cross* **105**, 1690–1728, doi:10.1017/S1816383123000413 (2023).

4 Compassion as a Timely and Timeless Moral Guide for Digital Innovation

Contemporary digital innovation is organised around optimisation, efficiency, scale, disruption, and growth. Technologies are judged by their capacity to increase engagement, accelerate processes, reduce friction, and expand market share. Success becomes legible through metrics: user acquisition, time on platform, throughput, and valuation. This logic rewards what can be measured and compared, privileging speed over reflection, scale over situated practice, and generalisability over context.

Ethical reflection is thereby reduced to risk management, focused on limiting liability while preserving competitive advantage. Less visible are harms that evade dashboards, such as diffuse impacts on marginalised communities, erosion of attention and trust, and entrenched asymmetries of power. These harms disperse across time and space, are externalised onto those least able to resist, or are normalised as the cost of doing business. When optimisation dominates, suffering is statistically diluted and morally abstracted. The challenge, then, is not better metrics but a different evaluative stance. Digital innovation requires not greater intelligence, but greater compassion.

When I teach digital ethics at the University of Sydney, the most consistent surprise among students is that there is more than one ethical theory. Many arrive believing that ethics simply means identifying the action that produces the greatest good for the greatest number. This is unsurprising: consequentialist reasoning dominates business and engineering curricula, where ethical reasoning (if it is taught at all) is often presented as a matter of calculating trade-offs.

When students then encounter deontological and virtue-ethical approaches, which sometimes yield contradictory judgements, the reaction is often confusion, anxiety, or even frustration. If respected ethical theories disagree, they ask, is moral action arbitrary? And how should one reason ethically when technologies reshape societies faster than norms or laws can stabilise? Such disagreement, however, signals not failure but the need for a constitutional orientation from which competing theories can be interpreted in light of lived vulnerability. What is required is not another theory, but a way to navigate among them. Compassion provides such an orientation without displacing their insights.

DOI: 10.4324/9781003614814-5

Digital technologies operate at scales and speeds that exceed assumptions embedded in traditional frameworks. Their malleability, generativity, and data-driven feedback loops enable rapid diffusion across domains, producing consequences that unfold through opaque, non-linear pathways. Unlike material goods, digital artefacts replicate at near-zero marginal cost and scale globally with ease, intensifying risks of monopolisation and exclusion. Network effects and platform lock-in consolidate advantage, limiting meaningful exit. As general-purpose infrastructures, digital technologies reshape multiple sectors while concentrating economic and political power among platform owners. Their transnational reach further allows them to exploit jurisdictional gaps, leading harms to accumulate without clear accountability. The result is diffuse, systemic externalities – surveillance, polarisation, environmental strain, dependency – that classical theories were not designed to anticipate.[1]

In such turbulence, no single "correct" theory is sufficient. What is needed is an orientation attuned to complexity and vulnerability. Compassion disciplines attention by directing focus toward those who bear digital innovation's costs and whose suffering is most easily obscured. It reorders moral salience, foregrounding differential vulnerability over aggregate performance, and thereby reshapes what counts as ethically relevant.

Compassion resurfaces in periods of transformation because it centres those most at risk. In digital contexts, vulnerability is structural, not exceptional. Agency is distributed across vast networks, obscuring accountability. Algorithms mediate attention, identity, and opportunity, often reinforcing inequality. Harms accumulate incrementally through designs optimised for engagement or extraction rather than well-being. Frameworks developed for stable institutions struggle to capture these dynamics.

Compassion does not offer a catch-all solution, but it does foreground the lived realities behind abstract metrics. It asks whether technologies reduce or intensify avoidable suffering, especially for those silenced, most exposed to harm, or constrained in agency. In doing so, it places vulnerability, relational responsibility, and care at the centre of ethical reflection on digital innovation, framing the analysis that follows.

The "Big Three" Ethical Traditions and Their Implications for Digital Innovation

Before turning to the three dominant traditions, it is important to clarify the purpose of this discussion. What follows is not a neutral survey or an attempt to rank frameworks. It proceeds diagnostically. Each tradition illuminates key moral aspects of digital technologies while rendering others less visible. Consequentialism highlights aggregate outcomes, deontology duties and rights, and virtue ethics character and moral formation. Yet in sociotechnical environments marked by scale, opacity, and asymmetrical power, these strengths can become blind spots. Distributed harm, slow violence, and structural vulnerability often fall between lenses. Viewed through a compassionate orientation, such blind spots become clearer. Compassion does not displace these traditions, but it does reframe them by asking whose suffering is backgrounded, whose voice is muted, and whose risk is normalised.

Consequentialism, deontology, and virtue ethics continue to shape reasoning across business, medicine, engineering, and law. Each offers a distinct account of moral action and valuable insights for digital design and governance. Yet globally scaled, algorithmically mediated systems strain their assumptions.[2] Compassion matters here not as a replacement, but as a way to address limits while preserving strengths. Table 4.1 provides an overview.

Table 4.1 The "Big Three" Ethical Traditions Reframed for Digital Innovation

Tradition	*Definition*	*Compassionate Reframing for Digital Innovation*
Consequentialism/ Utilitarianism	Evaluates actions by their outcomes, aiming to maximise aggregate benefit or minimise suffering. In digital innovation, it appears in cost–benefit analysis, optimisation metrics, growth targets, and claims that technologies make "most people better off".	Compassion shares the goal of reducing suffering but resists metric reductionism. At global scale, aggregate gains can mask patterned harm to vulnerable groups. Compassion insists that the distribution of harm matters as much as its total quantity, challenging trade-offs that justify predictable harm to some for measurable benefit to others.
Deontology	Grounds morality in duties, rights, and universal principles that constrain action regardless of outcomes. In digital innovation, it surfaces in privacy rights, anti-discrimination law, and regulatory mandates.	Compassion deepens duty by rooting it in attentiveness to lived vulnerability rather than abstract rule-following. In sociotechnical systems, formal compliance may satisfy legal standards while leaving structural injustice intact. Compassion demands that rights be interpreted in light of power asymmetries, constrained consent, and unequal exit options.

(*Continued*)

Table 4.1 (Continued)

Tradition	*Definition*	*Compassionate Reframing for Digital Innovation*
Virtue Ethics	Focuses on moral character and cultivated dispositions such as prudence, justice, and courage. In digital innovation, it appears in appeals to responsible leadership, ethical design, and professional norms.	Compassion grounds moral agency where rules are incomplete and consequences uncertain. In systems that reward speed and scale, it sustains discernment and relational responsibility. It counteracts exploitative design logics, such as attention extraction and manipulative nudging, and reorients institutional culture toward dignity rather than innovation alone.

Consequentialism judges actions by their outcomes, aiming to maximise benefit or minimise suffering. Bentham and Mill formalised this tradition as utilitarianism, which remains influential in policy analysis, cost–benefit evaluation, and corporate decision-making.[3] Compassion aligns readily with consequentialist thinking because it shares the aspiration to alleviate unnecessary suffering and foster human flourishing. Yet applying consequentialism to digital innovation exposes a deep problem. Cost–benefit analysis works only when consequences can be mapped, compared, and anticipated.

Digital technologies routinely defeat these conditions. The effects of engagement-maximising platforms, algorithmic classification, or data-driven labour markets unfold through long, complex, and multifactorial causal chains stretched across institutions and continents. Aggregated benefits often obscure structural inequities, and claims that an innovation makes "most people better off" can hide that those harmed are consistently those already vulnerable. Compassion challenges this moral arithmetic by insisting that the distribution of harm matters as much as its total quantity. It directs ethical attention to the lived experiences behind the numbers and resists the idea that predictable harm to some can be justified by gains for others. In this way, compassion preserves what is valuable in consequentialism while rejecting the reductionism that often accompanies it.

Deontology approaches ethics from the standpoint of duty rather than consequence. Kant gave this tradition its most influential articulation through the idea that moral rules must be universal, rational, and unconditional.[4] In digital innovation, deontological principles appear in privacy rights, anti-discrimination protections, and fairness requirements that constrain what organisations may and may not do.[5] Yet rules written for stable institutional environments can become brittle when applied to dynamic sociotechnical systems. Compliance may satisfy legal obligations while leaving substantive injustices intact. Duties interpreted abstractly can also lose sight of the people whose lives are shaped by digital infrastructures they neither meaningfully consented to nor can realistically opt out of.

Here, compassion provides a corrective by grounding duty in attentiveness to vulnerability. It requires that duties be interpreted in light of who may be harmed and how, rather than applied in isolation from social context. Compassion does not weaken the deontological commitment to dignity, autonomy, or respect; it strengthens it by insisting that moral principles remain responsive to lived circumstances.[6] While Kant imagined ethics as an exercise of pure reason, Schopenhauer later argued that reason alone cannot motivate genuine moral concern. Compassion, for him, is the only moral motive that crosses the boundary between self and other.[7] His critique does not reject duty, but deepens it by connecting moral rules to the recognition of shared suffering. This relational understanding of duty becomes especially important when digital technologies affect people unevenly and when those most affected are least able to influence their design.[6]

Virtue ethics offers a third foundation for ethical reasoning, grounded in the ancient Greek insight that morality is not only about rules or outcomes, but about character. Aristotle described virtues as habits of good judgement cultivated through practice, guided by reason and oriented toward human flourishing.[2] Compassion sits naturally within this tradition as a disposition to recognise and respond to suffering in ways that are proportionate, discerning, and grounded in care.

Virtue ethics is well suited to digital innovation because it acknowledges that rules will always be incomplete and that complex decisions require discernment rather than formulae.[8] Yet cultivating virtue is challenging in environments where institutional incentives reward haste, experimentation, and disruption. When organisations prioritise scale and novelty over responsibility, the conditions that support moral character can erode.

Compassion helps sustain virtue by orienting individuals and institutions toward relational responsibility. It counterbalances impulses that drive harmful design choices, such as exploiting attention, manipulative behavioural nudging, or externalising emotional and cognitive labour. Compassion, understood as a virtue, fosters a mode of moral agency that is iterative, reflective, and sensitive to context. Such qualities are essential for navigating the ethical dilemmas of sociotechnical change.

Although often framed as Western traditions, these theories sit within a broader moral landscape. Across cultures, compassion appears as foundational. In Confucian thought, *rén* denotes humane benevolence grounded in attentiveness to others and the cultivation of relational harmony. In Buddhist ethics, compassion is the highest form of moral wisdom, essential to recognising universal suffering and responding to the conditions that produce it; it is not merely an emotion but a disciplined practice. Hindu traditions emphasise *ahimsa*, the principle of non-harm that extends to all living beings. Islamic moral thought ties compassion to divine mercy, expressed through commitments to charity, justice, and mutual care. Christian traditions similarly place compassion at their centre through teachings on mercy, forgiveness, and neighbourly love. African humanist philosophies such as Ubuntu affirm that a person becomes a person through others, highlighting interdependence and shared responsibility for community well-being. While their vocabularies differ, these traditions converge on compassion as a moral orientation.[1]

This convergence matters because digital infrastructures span cultures and histories. Ethical reasoning confined to Western frameworks risks reproducing the asymmetries digital technologies amplify.[9] A compassion-oriented approach encourages plural, context-sensitive reflection.[1] It recognises historically shaped values and uneven global distributions of benefit and burden. It also demands attention to structural sources of suffering – from data extraction in low-bargaining contexts to outsourced content moderation and algorithmic systems reinforcing linguistic hierarchies. Ethical responsibility thus extends beyond avoiding intentional harm to redressing embedded asymmetries.

Compassion becomes a unifying foundation not by replacing the Big Three, but by drawing together their strengths. It shares consequentialism's concern for reducing suffering while resisting metric reductionism. It affirms deontological dignity while keeping duties responsive to context. And it echoes virtue ethics' focus on character while situating moral agency within institutions. It enables multiple ethical languages to remain distinct yet mutually intelligible. Amid the ethical complexity of sociotechnical change, this orientation offers a humane basis for reasoning attentive to power, vulnerability, and avoidable suffering.

Compassion does not eliminate dilemmas or supply formulaic solutions. Its role is prior. It shapes perception by centring lived vulnerability over abstract coherence, shifting deliberation from aggregate performance to those most exposed to harm. It recalibrates responsibility, demanding responsiveness proportionate to power. As an orientation, compassion restructures moral perception before guiding judgement. To see why this demand is so exacting, we turn to Schopenhauer.

Schopenhauer's Philosophy of Compassion

Arthur Schopenhauer (1788–1860) offers a powerful foundation for understanding compassion as an ethical orientation suited to periods of technological and social upheaval. Often caricatured as pessimistic and obscure, his work in fact articulates a rigorous and humane moral vision. He not only identifies compassion as morality's basis, but also explains why it arises, what it addresses, and how it enables navigation of a world structured by power, striving, and suffering. His system of thought spans the four major branches of philosophy: ontology (the study of being and its underlying nature), epistemology (the study of knowledge and how it is formed), ethics (the study of moral principles guiding action), and aesthetics (the study of beauty and taste). Together, these domains show that compassion is not sentimental excess but a response to fundamental conditions of existence – conditions digital innovation frequently intensifies by amplifying striving, mediation, and opacity.[8] Table 4.2 provides an overview.

Table 4.2 Schopenhauer's Philosophy and Compassionate Digital Innovation

Branch	*Schopenhauer's Key Insight*	*Implications for Compassionate Digital Innovation*
Ontology: The study of being and its underlying nature.	**Metaphysical Voluntarism:** Reality is driven by *Wille, that is,* a blind, restless striving manifesting as desire and craving. Because desire is never finally satisfied, suffering is intrinsic and shared.	If suffering is intrinsic to existence, digital technologies that amplify striving through engagement metrics, comparison architectures, and perpetual stimulation intensify vulnerability. Compassion becomes a counterweight to technologies that monetise restlessness, recognising shared vulnerability while demanding heightened responsibility from those who control systems.
Epistemology: The study of knowledge and how it is formed.	**Subjective Representationism:** The world is encountered as *Vorstellung, that is,* mediated representation shaped by cognitive-perceptual structures. We never access reality directly; insight into shared vulnerability arises through introspection into our own striving.	Digital technologies structure representation at scale, curating visibility and shaping moral salience. Compassion therefore requires epistemic humility: what is most visible is not necessarily what is most vulnerable. It resists algorithmic instrumentalisation and redirects attention to obscured and marginalised suffering.

(*Continued*)

Table 4.2 (Continued)

Branch	*Schopenhauer's Key Insight*	*Implications for Compassionate Digital Innovation*
Ethics: The study of moral principles guiding action.	**Compassionate Imperative:** Compassion, not abstract reason, grounds morality. Recognising another's suffering as continuous with one's own dissolves egoistic boundaries. The highest moral duty is thus to harm no one, and help everyone as far as possible.	In systems marked by diffused agency and asymmetrical power, responsibility cannot rest on intention or compliance alone. Compassion reframes it as responsiveness to avoidable suffering. The vulnerability principle demands proportional protection for the powerless and accountability for those with power.
Aesthetics: The study of beauty and contemplative experience.	**Aesthetic Release from Striving:** Art suspends servitude to *Wille*, enabling perception free from craving and instrumentalisation. Such moments cultivate humility, receptivity, and attentiveness.	Digital technologies optimised for speed and stimulation erode the contemplative capacities moral perception requires. Compassion depends on attentiveness, which requires environments allowing reflection. A compassionate ethic therefore calls for systems oriented toward contemplative appreciation rather than extraction.

My engagement with Schopenhauer began well before academic life, during my undergraduate years at the University of Stuttgart. I spent long days in the university library, moving between fascination with information systems and a lingering sense that, important as it was, the field did not address the deeper questions that occupied me. In the neighbourhood where I lived, a stern stone bust of Schopenhauer stood above a small theatre, a silent guardian of the area's artistic undercurrent. My uncle, an actor, spoke of him with near-devotional reverence, and many musicians and artists I knew shared that fascination. One weary afternoon, seeking respite from relational database theory, I wandered into the philosophy section and picked up *The World as Will and Representation.*[10] Reading it in the original German felt like entering a different intellectual universe that spoke directly to suffering, desire, beauty, and the human condition. I immersed myself in his works, returning to them as my research turned toward sociotechnical systems. This chapter introduces a thinker whose clarity and moral seriousness have shaped how I approach the dilemmas we study.

Schopenhauer's ontology, known as metaphysical voluntarism, begins with the claim that the world in its essence is *Wille* – literally "will", but with connotations such as desire, craving, or striving. Wille is a blind, restless force expressed in all beings as individual motives and longings. Human agency is not sovereign over this striving but caught within it. Because desire is never satisfied for long, striving is continuous and suffering structural. As one craving subsides, another emerges. Suffering, for Schopenhauer, is not an anomaly but an existential condition shared by all beings animated by *Wille*.[10]

This ontological backdrop renders compassion visible. If suffering is universal and rooted in the nature of existence, moral life begins with recognising intrinsic vulnerability. Here, Schopenhauer offers profound insight for digital innovation. Contemporary technologies do not merely reflect desire; they amplify and monetise its restlessness. Notifications, metrics, personalised feeds, and algorithmic incentives provoke new cravings and cultivate habits that resist rest. Systems designed to optimise engagement magnify the very forces Schopenhauer saw as the source of suffering. Where striving is ubiquitous, vulnerability is universal, and compassion becomes a necessary moral response.[8] Contemporary platforms intensify this condition. Architectures organised around engagement, visibility, and perpetual update convert attention into a scarce and contested resource. The result is not only distraction but also structurally reinforced dissatisfaction. In such environments, compassion becomes a counterweight to systems that magnify striving while obscuring its human costs.

Schopenhauer's epistemology complements this ontology. The world, he argues, is encountered not directly but as *Vorstellung* – typically translated as "representation" yet better understood as a performance of objects in the theatre of the mind. The term matters. It is not merely that things are presented to us; we also actively engage with them, shaping how they appear through our cognitive, affective, and perceptual capacities. Following Kant, Schopenhauer maintains that what we perceive is always mediated through the structures of mind, that is, our categories, concepts, intuitions, and forms of understanding. But he goes further by insisting that insight into the underlying *Wille* is gained not through empirical investigation but through introspection into our own desiring, striving, and suffering. Because the same *Wille* animates all beings, recognising our own vulnerability becomes a window into others'. The boundary between self and other softens; another's suffering appears as akin to our own.[7,10]

This epistemology carries renewed significance for digital innovation. Platforms curate visibility, frame salience, and filter what appears real according to institutional and commercial logics largely opaque to users. Representation is never neutral. Digital technologies structure moral perception, shaping whose suffering is seen, ignored, trivialised, or amplified. Compassion therefore requires epistemic humility, a recognition that what is most visible is not necessarily what is most vulnerable. It tempers overconfidence in algorithmic neutrality and redirects attention to obscured realities.

Within this interplay of ontology and epistemology, Schopenhauer grounds his ethic of compassion. If all beings manifest the same striving, compassion arises when another's suffering is perceived as continuous with one's own. It is not calculation or duty but immediate participation in another's pain, a moment in which the veil between self and other lifts. From this follows his compassionate imperative: *hurt nobody, and help everybody as much as you can.*[7,8] This imperative is simple yet demanding. It prohibits treating any person as a means to an end. It requires proactive alleviation of suffering proportional to one's capacity. And it rejects the moral permissibility of harming a few for the supposed benefit of many.

Compassion, however, is more than non-harm. It is a sustained orientation of attentiveness and responsiveness. Schopenhauer contrasts this with the golden rule. Treating others as we wish to be treated risks projecting our standpoint onto them. Instead, we must treat others as *they* would wish to be treated, which demands sensitivity to contextual, cultural, and relational differences.[8] In digital innovation, this distinction matters. It guards against paternalistic design, against assumptions that engineers or policymakers know what is best for users, and against solutions that reproduce harm by universalising privileged perspectives. It aligns with feminist ethics of care, which emphasise attentiveness, relationality, and responsiveness to concrete needs, and with Buddhist and Hindu traditions that foreground compassion, non-violence, and interdependence.[8]

Schopenhauer was among the first Western philosophers to engage seriously with these traditions, noting resonance between his account of suffering and the Four Noble Truths of Buddhism: that life entails suffering; that suffering arises from craving; that liberation requires the cessation of craving; and that ethical and contemplative practice can reduce suffering for oneself and others.[8] This cross-cultural convergence strengthens the case for compassion as a universal moral orientation grounded in shared vulnerability rather than cultural particularity.

Still, Schopenhauer's compassionate imperative has limits, and nor was it intended to be universal or exhaustive. He distinguished it from Kant's categorical imperative on the grounds that elevating one's own moral views to a universal law is itself an act of egoism and therefore morally hollow.[7] This humility is a strength. Recognising the limits of his formulation is essential for adapting it to digital innovation. His emphasis on universal suffering, while morally expansive, does not fully address the patterned, structural, and historically produced dimensions of vulnerability shaping contemporary societies.

Not all suffering is equal. Some suffering arises from existential striving, but much stems from entrenched injustice. Colonial dispossession, racialised violence, extractive capitalism, gendered exploitation, and cumulative political and economic inequities create suffering that is unevenly distributed and disproportionately borne by those with the least power, who did not create the structures they face.[11] Schopenhauer's injunction to "harm no one" does not explicitly address cases in which benefiting from historically produced injustice already constitutes harm, or in which redressing injustice requires those with inherited advantage to relinquish privilege. Without such recognition, compassion risks sliding into a conservative ethic that prioritises conflict avoidance over the redistribution of power.

This is why the vulnerability principle matters. While Schopenhauer foregrounds universal vulnerability, contemporary ethical life requires attention to differential vulnerability.[6] A compassionate orientation must be proportionately gentle toward those who lack power and proportionately demanding toward those who wield it. Designers, platform owners, policymakers, and engineers therefore bear heightened responsibility not only because they can cause harm but because they occupy upstream positions of infrastructural influence. Conversely, those exposed to systemic disadvantage require not merely "neutrality", but downstream protection and redress. Compassion, understood in this way, is not a feel-good sentiment but a radical ethical stance.[11] It calls for a rebalancing of burdens, recognising that neutrality in the face of structural injustice effectively sides with power.[6]

The vulnerability principle becomes concrete from the standpoint of users. Consider a teenager navigating social media environments calibrated for visibility and comparison, where metrics of popularity shape self-understanding at formative stages of identity. Or a gig worker whose livelihood depends on opaque algorithmic scoring systems that rank, allocate, and discipline without meaningful recourse. Or an underage girl targeted by deepfakes, confronting reputational harm amplified by viral circulation and platform persistence. In each case, vulnerability is not incidental but structurally produced. Exposure, dependency, and limited capacity to contest platform decisions define the lived experience of participation. Compassion, as an orientation, foregrounds these asymmetries and treats them as morally salient rather than peripheral side effects of innovation.

A contrasting vantage point is that of designers and product owners. Here, decisions are shaped by incentive structures, growth targets, and dashboards that reward engagement, retention, and monetisation. Responsibility is often diffused across teams, supply chains, and hierarchical layers, making the source of harm difficult to localise.[12] Yet power remains asymmetrically concentrated: design choices configure the field within which others act. Compassion requires those with structural influence to interrogate not only outcomes, but also the differentiated risks their systems impose.

Schopenhauer's aesthetic theory deepens these commitments. He argues that aesthetic contemplation temporarily suspends our servitude to Wille and reveals the world without the distortions of desire, competition, or instrumentalisation. Music, art, and beauty offer glimpses of reality not governed by striving but by perceptual stillness. These moments cultivate capacities essential for compassion: receptivity, humility, attentiveness, and the ability to see others not as means but as beings with inner worlds.[10] In digital environments saturated with stimuli designed to intensify craving, such capacities are eroded. Constant distraction diminishes the contemplative space required for moral perception.

Through Schopenhauer's lens, well-being becomes not a matter of individual self-control but a structural precondition for ethical life. Technologies that undermine contemplative attention corrode the conditions under which compassion can flourish.[8] Just as ultra-processed foods, driven by commercial interests and government inaction, displace the conditions for physical health, contemporary social media's supply of ultra-processed speech crowds out the reflective modes of communication on which moral discernment depends. The challenge is not merely individual failure to regulate media diets, but an informational environment optimised to erode the contemplative and relational capacities required for compassion to take root.[13]

Schopenhauer's reflections on aesthetics also offer a pointed critique of contemporary digital design. He understood beauty as a brief release from restless striving, a suspension of craving that makes contemplative perception possible. Yet today's digital spaces often resemble architectural brutalism: efficient, imposing, materially honest in their emphasis on functionality and throughput. Brutalism has its own aesthetic – inexplicably to me, many find it beautiful – yet few would wish to inhabit a city composed entirely of it. Flourishing environments require contrast: ornament, softness, neoclassical curves, art deco flourishes, or simply room to breathe.

Digital infrastructures are rarely designed with such plurality in mind. They commodify expression, connection, and intimacy into channels for data extraction, leaving little space for experiences valued for their own sake or conducive to stillness. A compassionate ethic challenges this narrowing of aesthetic possibility by insisting that humane engagement depends on environments that support attentiveness, reflection, and non-instrumental encounter. Digital spaces could host such experiences, but too often they intensify striving rather than quiet it, offering efficiency without grace and saturation without depth.[8]

Across these philosophical pillars, Schopenhauer develops a unified account of moral life grounded in shared vulnerability and the imperative to alleviate suffering wherever possible. His philosophy is pertinent for digital contexts in which harms are inequitably distributed, causality is opaque, and responsibility is diffused. Traditional ethical frameworks tie responsibility to intention or control, yet digital innovations routinely exceed both.[13] Algorithms shape behaviour without conscious direction, and platform dynamics generate harms through feedback loops no single designer foresaw. The compassionate imperative reframes responsibility not as blame but as responsiveness. It asks not "who caused this?" but "how can we lessen the suffering before us?" Responsibility becomes proportional to capacity, and inaction acquires moral weight when it permits avoidable harm to persist.

Schopenhauer's thought is therefore neither antiquated nor abstract. It offers a vocabulary for understanding why digital innovation poses acute ethical challenges and why established traditions struggle to navigate them. His emphasis on vulnerability, relationality, and cognitive limits anchors ethical life in humility rather than control. His critique of egoism exposes how digital infrastructures privilege the powerful over the vulnerable. His ethic of compassion demands a reversal of this logic and provides a guide for evaluating technologies and the institutions that design and deploy them. It does not resolve the dilemmas of sociotechnical change, but equips us to confront them with care. It insists that the measure of innovation is not novelty, efficiency, or profitability, but its contribution to reducing suffering and enabling the flourishing of those most at risk. As we turn to the applied chapters on social media, Web3 governance, and AI companions, this ethic serves as the conceptual anchor shaping the moral horizon against which digital futures must be judged.

Letting Compassion be the Guide

If compassion is a constitutional orientation, the task ahead is not its performative endorsement, but its institutionalisation. How can such an orientation be embedded in design practices and product architectures, operationalised in governance arrangements and accountability mechanisms, and translated into regulatory oversight? How might it reshape incentives that privilege scale, speed, and engagement over situated well-being?

The ethical dilemmas of digital innovation arise in concrete sociotechnical settings where design choices, institutional incentives, and human vulnerabilities collide. As Chapter 3 showed, these domains are not neutral infrastructures but value-laden environments in which power, legitimacy, and responsibility are continually renegotiated. Compassion clarifies why these tensions persist. Digital technologies intensify striving, structure representation, and erode the contemplative capacities needed for ethical discernment. They also obscure the suffering they help produce. What is required now is to translate this philosophical foundation into analysis of the technologies that most directly shape contemporary life.

Compassion shifts ethical attention from optimisation to vulnerability. Rather than asking only what technologies can do or what aggregate outcomes they yield, it asks how their design, governance, and use shape the lives of those most at risk of harm. In environments marked by distributed agency, opacity, and rapid change, harms often accumulate through feedback loops that reward craving, through representational architectures that narrow perception, and through governance arrangements that externalise risk. Compassion challenges the assumption that such harms are inevitable by-products of progress. It asks who suffers, why, and how avoidable suffering might be reduced. Each dilemma becomes not merely a trade-off but a site of moral contestation and an opportunity to rethink power, responsibility, and accountability.

Letting compassion guide digital innovation cannot rely on individual virtue alone. Personal responsibility, while important, is insufficient in systems governed by scale, automation, and competitive pressure. Compassion must be institutionalised: embedded in governance structures, review processes, incentive systems, and design protocols that shape everyday decisions. This requires translating attentiveness to vulnerability into operational constraints, placing limits on extractive data practices, safeguards against asymmetrical risk transfer, and accountability mechanisms proportionate to power. It also entails commitments to redress, transparency, and participatory voice, particularly for those most exposed to harm. Without such embodiment, compassion risks devolving into empty rhetoric invoked in mission statements yet absent from the architectures and economic logics that organise digital life.

The chapters that follow examine three technologies that exemplify these dynamics: social media, Web3, and AI companions. Social media reveals how architectures of representation mould collective attention and conflict. Web3 exposes how institutional design embeds competing values of autonomy and coordination, often reproducing inequalities that decentralisation promises to disrupt. AI companions extend these tensions into the intimate sphere, where vulnerability intensifies and the boundary between care and manipulation becomes fraught. Across all three domains, compassion serves as the evaluative thread linking philosophical groundwork to practical analysis.

What follows is not a search for ideal solutions but an inquiry into what becomes possible when compassion guides navigation of ethical dilemmas that are ongoing, relational, and historically conditioned rather than discrete problems to be solved. The task is to explore how compassion can inform interventions that expand human flourishing within systems increasingly governed by algorithmic and economic logics. In this sense, compassion functions not only as an ethical orientation but as a method for reimagining digital futures, recognising that today's choices will harden into the infrastructures future generations will inhabit.

References

1 Ciriello, R. F., Mathiassen, L., Risius, M., Cheong, M., Scheepers, H., & Vaast, E. Compassionate Digital Innovation: A Pluralistic Perspective and Research Agenda. *Information Systems Journal* **36**, 364–385, doi:10.1111/isj.70003 (2025).

2 Gal, U., Hansen, S. & Lee, A. S. Research perspectives: toward theoretical rigor in ethical analysis: the case of algorithmic decision-making systems. *Journal of the Association for Information Systems* **23**, 1634–1661, doi:10.17705/1jais.00784 (2022).

3 Mill, J. S. *Utilitarianism.* (Oxford University Press, 1863).

4 Kant, I. *Grundlegung zur Metaphysik der Sitten.* (J.F. Hartknoch, Riga, 1785).

5 Chatterjee, S., Sarker, S. & Fuller, M. A. A deontological approach to designing ethical collaboration. *Journal of the Association for Information Systems* **10**, 138–169, doi:10.17705/1jais.00190 (2009).

6 Ciriello, R. F., Chen, A. Y. & Rubinsztein, Z. A. Compassionate AI design, governance, and use. *IEEE Transactions on Technology and Society* **6**, 1–6, doi:10.1109/TTS.2025.3538125 (2025).

7 Schopenhauer, A. *Über die Grundlage der Moral.* (Felix Meiner Verlag, 1840).

8 Ciriello, R. F. Dear Schopenhauer: how can IS artefacts help us flourish? *Wille, Vorstellung,* compassion, and beauty in social media. *Scandinavian Journal of Information Systems* **37**, 317–354, https://aisel.aisnet.org/sjis/vol37/iss1/9/ (2025).

9 Atari, M., Xue, M., Park, P., Blasi, D. & Henrich, J. Which humans?, doi:10.31234/osf.io/5b26t (2023).

10 Schopenhauer, A. *Die welt als wille und vorstellung.* (Brockhaus, 1859).

11 Ciriello, R. F. & Mindel, V. Techno-collectivism: scaling libertarian socialist governance of digital infrastructure. *IEEE Technology and Society Magazine*, 1–15, doi:10.1109/MTS.2025.3605115 (2025).

12 Birsch, D. Moral responsibility for harm caused by computer system failures. *Ethics and Information Technology* **6**, 233–245, doi:10.1007/s10676-005-5609-5 (2004).

13 Ciriello, R. F., Gal, U., Hannon, O. & Thatcher, J. Responsible social media use: how user characteristics shape the actualisation of ambiguous affordances. *European Journal of Information Systems* **34**, 799–821, doi:10.1080/0960085X.2024.2444249 (2025).

Part II

Critique

5 Compassionate Social Media Use

From their early 2000s origins as quirky college networks and hobbyist forums, social media platforms evolved into indispensable infrastructures of public life. This shift from playful gadget to extractive platform crystallises the ethical tensions of sociotechnical change. What once promised connection and creativity now structures visibility, identity, and political discourse for billions, magnifying tensions that earlier technologies contained at a smaller scale.[1] As surveillance advertising displaced community norms, users entered environments where freedom and harm, intimacy and surveillance, and solidarity and hostility coexist in volatile combinations.[2]

The history of social media is thus one of value reconfiguration: inherited norms of expression and responsibility were stretched or hollowed out as corporate platforms enclosed everyday communication. As Chapter 2 argued, social media's volatility arises not merely from misuse but from ambiguous affordances that shape its routine enactment. Even so, personal responsibility still matters.

Today, social media has become a core infrastructure of public life. It enables connection and mobilisation at scale, yet it also fuels misinformation, polarisation, social comparison, and declining trust.[3,4] The rapid spread of high-volume, low-quality generative AI content ("AI slop") intensifies this dynamic by saturating feeds with indistinguishable synthetic material. As authenticity becomes harder to verify, users may treat all content as potentially manipulated, echoing American novelist Danielewski's warning of a "climate of reduced expectations" in *House of Leaves*, where visual representation loses evidentiary force. In French philosopher Baudrillard's terms, digital media drifts toward *simulacra*, where representations no longer refer to an underlying reality but circulate as self-contained signs. Practices such as deliberately "messy" posting among Gen Z or retreat into private, ephemeral spaces (e.g., encrypted or disappearing messaging platforms) can be understood as adaptive responses to declining trust in digital media.

These tensions stem from corporate platforms that host civic exchange while optimising for engagement and commercial value.[5] Accordingly, scholarly and policy debates emphasise regulation and platform governance, calling

DOI: 10.4324/9781003614814-7

for algorithmic transparency, privacy protections, and democratic oversight. Yet platforms operate not only through design choices but through billions of user micro-actions. Scrolling, liking, sharing, reacting, and even remaining silent collectively shape the affective atmosphere of digital life. Even powerful systems rely on habitual conformity.[1]

Recognising this interdependence creates space to consider personal responsibility without reverting to neoliberal victim-blaming. Corporations have long mobilised narratives of individual responsibility to deflect accountability for structural harms. The "carbon footprint" reframed ecological crisis as consumer choice.[6] "Jaywalking" campaigns recast pedestrian deaths as behavioural failures rather than products of car-centric planning.[7] In the infamously misrepresented McDonald's coffee case, a 79-year-old woman suffered third-degree burns after prior complaints, yet corporate lawyers constructed a caricature to delegitimise consumer rights as "frivolous lawsuits".[8] Such examples justify scepticism toward individualisation. Yet rejecting ideological blame does not require rejecting agency. It requires distinguishing between responsibility that shields power and responsibility that enables ethical action within constraint.

Czech political dissident turned president Václav Havel's *The Power of the Powerless* helps clarify this distinction.[9] Analysing everyday life in totalitarian Czechoslovakia, Havel argued that oppressive systems endure not only through central authority but through routine conformity. The grocer who displays the party slogan does so not from conviction but from habit and fear, and the system persists because millions act similarly every day. His claim was not that individuals are to blame, but that even modest refusals can unsettle domination.

Today's digital infrastructures differ from the post-totalitarian context Havel described, yet his structural insight remains instructive. Contemporary platforms are governed not by party-states but by capitalist market logics that generate comparable pressures toward behavioural conformity. Their malleability, generativity, data-driven feedback loops, zero marginal costs, network effects, and global reach embed them as general-purpose infrastructures across social life.[10]

These conditions create infrastructures that thrive on predictable user behaviour: steady attention, routinised emotional labour, and the reproduction of patterns that optimise engagement. Their influence is vast precisely because they embed themselves as general-purpose technologies across multiple social, cultural, and political life spheres. Yet their power is not absolute. If users interrupt predictable cycles by refusing to amplify outrage, resisting impulsive sharing, or disrupting escalatory interactions, the rhythms of these systems would shift, even if only partially.

The analogy to post-totalitarianism need not be literal to be instructive. Both systems stabilise through diffuse conformity, and both reveal how everyday refusals expose fissures in seemingly unassailable orders. Users are not

Table 7.1 Four Societal Currents Shaping the Rise of AI Companions

Societal Current	*Description*	*Contribution to the Rise of AI Companions*
The Loneliness Epidemic	Rising chronic isolation across age groups, especially youth. Greater mobility, weakened communities, flexible labour, and reconfigured families have expanded freedom while loosening durable ties. Social media increased visibility without delivering presence.	AI companions emerge as scalable substitutes for scarce intimacy. They provide always-available attention and simulated presence where human connection feels fragile. Loneliness shifts companionship from luxury to perceived necessity.
Neoliberal Marketisation of Social Life	Decades of market-centric governance hollowed out public institutions and reframed social goods as commodities. Platforms enclosed communication and monetised belonging through surveillance-based models. Social support increasingly flows through commercial infrastructures.	AI companions extend the commodification of intimacy. Emotional support becomes a subscription service shaped by retention, monetisation, and data extraction incentives. Care is industrialised by engagement logics.
Concentration of Tech Power and Capital	Economic and political power has consolidated in a handful of firms whose valuations depend on AI expansion. Massive investment in compute and proprietary models pressures firms to identify scalable, habit-forming applications.	AI companions offer high-retention, deeply personal, globally scalable products. Emotional dependency generates durable revenue and behavioural data, expanding infrastructural control over intimate domains.
Geopolitical AI Competition	Intensifying rivalry, particularly between the US and China, frames AI as strategic infrastructure. Compute capacity, semiconductor supply chains, and model dominance are treated as geopolitical assets.	AI companions become channels of soft power, persuasion, and surveillance embedded in daily life. Personalised influence at scale complicates regulation and heightens unity–diversity tensions across moral and national boundaries.

revenue farms. Social support systems were reframed as conditional privileges. In this vacuum, private platforms quietly installed themselves as the digital infrastructure of everyday life.[10] Social media promised to make the world more open and connected but instead enclosed connection, converting the human need for belonging into a multi-billion-dollar surveillance industry. Every like and share rests on extracting attention, behavioural data, and emotional vulnerability. Code, models, and data sit behind proprietary walls guarded by billionaire oligarchs who consolidate political power at scales democracies never anticipated.[11,12] AI companions inherit this lineage. They promise frictionless intimacy and judgement-free understanding, yet remain products shaped by the same extractive logic of engagement, retention, and monetisation. Commodified intimacy intensifies the autonomy–coordination dilemma, while harms generated through engagement-maximising architectures later appear as "unintended consequences".[13] How often does history need to rhyme before foreseeable, recurring consequences can no longer be excused as "unintended" by default?

The third force is the *concentration of economic and political power* in the tech sector. Markets now move not on productivity or human flourishing but on the projected earnings of a handful of firms whose valuations depend on speculative expectations about AI. Many billions of dollars flow into data centres that consume staggering amounts of energy, water, and rare-earth metals. No one has yet found a sustainable revenue model to cover the operational burn. But AI companions promise a lucrative lifeline: an endlessly scalable emotional service on which billions can literally depend, deeply personal, habit-forming, and hard to regulate. This dynamic is particularly explosive as the industry moves to adopt targeted advertising – the business model that made social media so enormously profitable and harmful – this time in more sensitive contexts where persuasion is likely amplified.[14] In this sense, AI companions become the perfect commodity in an economy that has commercialised nearly every other domain of human life. When a small number of firms set the infrastructural rules of intimacy, autonomy narrows to consumer choice while coordination concentrates at the top, locking present design decisions into long-term trajectories.

The fourth force is *geopolitical dispersion*. US dominance is fading; China's technological capacity now rivals or exceeds Western capabilities in key areas. A global AI arms race is underway, driven by the belief that whoever controls the most advanced models will shape economic, military, and political futures. This new Cold War plays out not through nuclear brinkmanship but through compute capacity, semiconductor supply chains, and influence campaigns embedded in everyday consumer apps. AI companions, with their millions of intimate conversations, become another frontier through which persuasion, surveillance, and ideological drift can be exercised. Personalised persuasion embedded in global platforms strains unity–diversity across moral worlds, while national oversight struggles to coordinate systems whose infrastructures exceed territorial jurisdiction.

Against this backdrop, AI companions have entered mainstream consciousness with breathtaking speed. ChatGPT reached one hundred million users in two months, unprecedented in the history of technology adoption.[15] And with speed came emotional intensity. What began as playful experimentation quickly hardened into attachment. People explored the technology for causally responsible for structural harm, yet they remain ethically implicated in its reproduction.

This interdependence has two implications for a compassionate orientation. First, individual behaviour is relational. Every click and silence influences narrative visibility and the circulation of public emotions within an attention economy. Second, responsibility concerns how actions affect others in environments shaped by corporate incentives and algorithmic filtering. Platform architectures encode commercial priorities; user practices encode values. Recognising agency does not deny manipulation or institutional failure. It acknowledges that agency endures and can be oriented toward compassion.

Responsibility spans legal, moral, and social dimensions.[11] Legally, individuals are accountable when they voluntarily contribute to harmful outcomes within their control. Sociotechnical systems complicate this by diffusing agency and obscuring causality. Morally, responsibility concerns judgement, integrity, and foreseeable consequences, shaped by evolving norms and the court of public opinion. Social responsibility attaches to roles, whose obligations often collide in the collapsed, multi-audience spaces of social media. Responsibility thus entails navigating overlapping expectations in environments where actions scale rapidly, roles blur, and consequences intensify.[1]

This perspective resonates with contemporary digital Luddism visible in hacktivism, digital minimalism movements, returns to "dumbphones", and critiques of addictive design. Contrary to its portrayal in popular discourse, Luddism did not reject technology per se; it resisted exploitation and oppression entrenched through technology. Its caricature as technophobia reflects corporate narrative capture and state repression.[12,13] Today's digital Luddism similarly targets concentrated power rather than tools. It pursues resistance to harmful infrastructures, removal of systems resistant to reform, and development of public-interest alternatives.[14] These strategies require collective action but depend on individual practices of refusal: withdrawal from manipulative patterns, disruptions of harmful interactions, and cultivation of more humane modes of participation. Compassion reframes responsibility not as a private disposition but as a micro-political contribution to systemic reform.

Still, the relation between structure and agency remains delicate. Platforms are sustained by billion-dollar business models optimised for behavioural extraction and weakly constrained by regulation. Against this stand the small but meaningful choices individuals can enact. A compassionate orientation never blames individuals for systemic failures, yet it acknowledges that personal responsibility remains consequential. Responsible use becomes a way of living deliberately within infrastructures that discourage deliberation. It echoes Havel's call to live in truth within systems that rely on routine compliance.

Compassionate social media use is therefore both moral and political. It concerns how individuals comport themselves in a noisy public sphere, respond to others' vulnerabilities, and manage their own. Such habits shape the conditions of public reason. Regulation, redesign, and reform are indispensable, yet they cannot reach every micro-interaction through which digital life unfolds. Conscientiousness, tolerance, and compassion can alter the moral texture of online spaces in ways institutional interventions cannot replicate.

For this reason, the chapter foregrounds the individual not as a site of blame but as a locus of agency under constraint. It asks how everyday engagement can become intentional and ethically attuned. It examines virtues such as conscientiousness, moral tolerance, and compassion as practical capacities within environments built for distraction and conflict. Behavioural patterns accumulate to shape the affective atmosphere of public discourse. Responsible use thus emerges as a collective possibility grounded in the insight that small acts of refusal can reshape the emotional and material conditions of society.

How Social Media Harms Emerge and Why Personality Still Matters

Free expression is central to human dignity because it enables people to articulate who they are, interpret their experiences, and participate in social life. It is also foundational to democracy: it allows people to contest injustice, coordinate political action, and scrutinise power. Yet today the legitimacy of expression is mediated largely by corporate platforms that determine which voices gain visibility and which are buried. This concentration of discursive authority is structural rather than incidental. Platforms sort, rank, and moderate speech according to engagement metrics, risk management, and advertiser expectations, exercising institutional power over billions. The ethical stakes of this configuration demand sustained vigilance.

These stakes become clearer when recognising the two structural tensions that define social media governance. The first is profitability versus well-being. Advertising-based models monetise attention, emotion, and behavioural data. They reward content that provokes outrage or fuels habitual scrolling even when such content corrodes psychological health, undermines civic trust, and distorts public discourse. Profitability requires maximising engagement; well-being requires curbing it. This contradiction is not an aberration but the business model's organising logic.

The second tension concerns freedom versus safety. People value freedom of expression but also freedom from harassment, manipulation, and harm. The openness that enables participation also enables abuse; efforts to reduce harm necessarily constrain some forms of expression. No governance model resolves this dilemma because the infrastructures themselves are engineered for speed, virality, and scale.

Users encounter these tensions through platform affordances that structure everyday interaction. Volatile spreadability enables mass mobilisation but also cascading harm.[1] The livestream of the Christchurch massacre showed how violent material can circulate globally within minutes despite automated detection.[15] Harassment campaigns – from the misogynistic online harassment campaign #Gamergate to coordinated pile-ons against journalists and activists – demonstrate how rapid diffusion concentrates aggression.[16]

Algorithmic targetability reinforces communities yet enables manipulation and ideological enclosure.[1] The Cambridge Analytica scandal illustrated how psychographic targeting could be weaponised for electoral influence.[4] YouTube's recommender system has nudged users toward extremist content through sequences of adjacent recommendations that encourage gradual ideological drift.[17]

Permanent immutability preserves records for accountability but allows harmful material to persist long after context fades.[1] Videos of police violence can support justice; youthful mistakes or intimate content can be recirculated indefinitely. These affordances are morally ambivalent. They enable solidarity and collective action while facilitating harassment and emotional exploitation. They are relational action possibilities enacted under engagement incentives.[1]

Understanding why individuals respond differently requires attention not only to personality traits but also to orientations enacted under constraint. Two dispositions are especially salient because they can be cultivated as attentional disciplines: conscientiousness and moral tolerance.

Conscientiousness reflects a person's capacity to regulate impulses, consider consequences, and take responsibility for their contributions. High conscientiousness introduces deliberation into digital interaction: users verify information, pause before posting, and resist reactive escalation. It enables what might be termed compassionate friction, slowing the spread of harm within systems optimised for speed and visibility. Low conscientiousness, by contrast, fuels impulsive engagement, spectacle, and emotionally charged reactions that sustain outrage economies and intensify harm.[1]

Moral tolerance captures one's willingness to engage differing values without hostility. High tolerance supports humility, perspective-taking, and sustained constructive engagement amid disagreement, thereby moderating polarisation in environments structured by algorithmic targeting. Low tolerance, by contrast, frames difference as a moral threat, fostering rigidity, suspicion, and confrontational engagement that legitimises ideological enclosure.[1]

These traits do not operate in isolation. They interact with platform affordances to shape how individuals behave in sociotechnical environments that reward speed, emotion, and constant visibility[1]. Dispositions and infrastructures thus co-produce patterns of conduct, giving rise to four recurring, though fluid, archetypes that shape compassionate or harmful engagement.

The *narcissist*, low in both conscientiousness and tolerance, is impulsive and self-centred, exploiting visibility for attention or dominance. Outrage-seeking influencers, users sharing decontextualised content, or those inciting harassment for entertainment exemplify this orientation. The *dogmatist*, high in conscientiousness but low in tolerance, combines discipline with rigidity, using rules to police ideological boundaries. Coordinated call-outs, purity tests, and moderators weaponising community standards illustrate how certainty becomes punitive. The *nihilist*, low in conscientiousness but high in tolerance, is marked by passivity and disengagement. Silent bystanders or unchallenged misinformation in private groups show how passive tolerance enables harm to spread. The *altruist*, high in both traits, combines self-regulation with openness, engaging in fact-checking, de-escalation, and support for vulnerable users, often at personal cost in environments that reward antagonism.[1]

Figure 5.1 visualises these archetypes along the axes of conscientiousness and moral tolerance. It offers a simplified analytical map rather than a fixed typology, enabling designers and policymakers to assess how behavioural orientations interact with platform incentives and where interventions (such as adding friction, slowing amplification, or supporting constructive engagement) may shift participation toward more compassionate forms.

These archetypes are heuristics, not fixed identities. People shift among them depending on mood, context, and norms. Compassionate adolescents may become dogmatic in political disputes; ordinarily conscientious adults may share impulsively during crises. Character in sociotechnical settings is situational and relational, shaped by dispositions as well as emotional pressures, cultural narratives, and engagement incentives.[1] At the same time, these

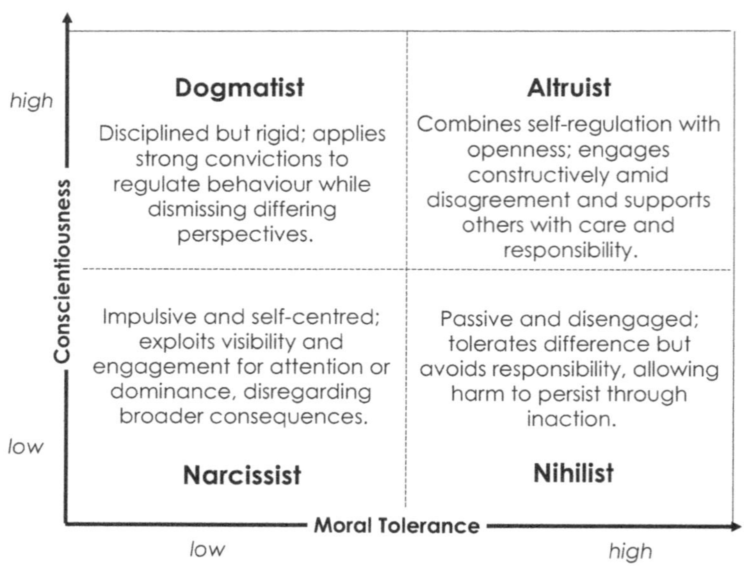

Figure 5.1 Character Dispositions and Personality Archetypes in Social Media Use.[1]

archetypes simplify complex platform dynamics, making behaviour easier for designers and policymakers to understand. They offer a lens for identifying how systems might slow escalation and support more compassionate participation. Compassionate use is therefore not simply virtuous expression but disciplined reorientation within extractive architectures. It means refusing to feed outrage economies, interrupting frictionless escalation, and responding proportionately to vulnerability even when systems reward the opposite.

Character expression is further shaped by sociotechnical conditions. Individuals display different facets of themselves with friends, colleagues, or family, and platforms amplify some dispositions while muting others.[1] Instagram's aesthetic norms can intensify comparison and envy, particularly among young women exposed to beauty filters that alter facial structure and skin tone.[18] TikTok's algorithm can amplify emotionally charged content, contributing to body dysmorphia and gendered hostility.[19] Recommender systems on YouTube, Instagram Reels, and TikTok can normalise reactive consumption, encouraging dogmatic or nihilistic tendencies even among otherwise responsible users.[1] Broader cultural narratives – political grievance, mistrust, economic precarity, status anxiety – further shape interpretation. Emotional proximity to harm may intensify reactivity, group solidarity may license cruelty, and silence may reflect fear rather than indifference. Interventions focused solely on individuals miss this conditioning. Yet structural constraint does not eliminate agency. It does clarify that compassion must operate across multiple spheres.

Within the personal sphere, compassion becomes a difficult but necessary mode of counter-conduct. Compassionate users often face goodwill abuse: their patience is exploited, their restraint used against them, and their refusal to retaliate becomes a reason to target them further. This is common in polarised environments where bad-faith actors reward escalation.[1] Compassionate users shoulder disproportionate emotional burdens not because compassion is weak but because hostile environments punish conscientious behaviour.

Compassionate practice should therefore not become a mechanism through which responsible users shoulder the weight of systemic dysfunction. Even so, it can alter the affective atmosphere of digital life. Declining to share inflammatory content slows cycles of spreadability. Verifying claims before reposting interrupts misinformation flows in private groups. Counter-speech weakens mob momentum. Refusing to reduce opponents to identity avatars disrupts polarisation, especially where emotionally charged snippets are recirculated for engagement. Each act introduces friction into systems built for amplification.

Systems that depend on predictable hostility are vulnerable to small disruptions. Compassion thus operates as a modest yet potent resistance. It creates space for dignity and reflection and unsettles the behavioural predictability on which algorithmic governance relies.

Social Media as a Paradigmatic Arena of Ethical Dilemmas

Social media concentrates the ethical dilemmas outlined in Chapter 3 because its infrastructures magnify value conflicts that already shape civic life. To integrate the chapter's dialectical foundation more fully, these tensions should be understood as mutually constitutive rather than as problems awaiting optimal balance.

Autonomy and coordination, progress, unity and diversity, intention and outcome are not variables on a slider but dynamic oppositions that generate and reshape one another over time. Efforts to suppress harmful expression, for example, often intensify claims of censorship, while unrestrained freedom predictably produces demands for regulation.[5] Likewise, the gap between intention and outcome can be understood as a dialectical misalignment between micro-level intent and macro-level effect, where individually reasonable actions aggregate into structurally predictable harm. Making these dynamics explicit clarifies that social media is not merely a case study but a live demonstration of enduring tensions.

Unity versus diversity forms the emotional core of the contemporary crisis. Platforms collapse contextual boundaries, forcing moral worlds that might otherwise coexist at respectful distance into direct confrontation.[1] This can foster learning, but it often fuels polarisation, as what seems self-evident in one community appears incomprehensible or offensive in another. Algorithmic curation intensifies this strain by amplifying affinity clusters while exposing users to unfamiliar norms without the interpretive scaffolding of offline life. The result is simultaneous enclosure and overexposure. The public sphere becomes broader, thinner, and more combustible. Unity erodes as shared ground dissolves; diversity becomes fraught as mutual recognition weakens. As Chapter 3 argued, no stable equilibrium resolves this dilemma. Social media simply renders its stakes more visible.

The tension of intention versus outcome is equally defining. Social media interactions scale unpredictably, and well-intended actions often generate harmful collective effects. A user may share information to help yet amplify misinformation; a joke may become a vector for harassment; a solidarity post may expose vulnerable individuals to scrutiny.[1] Platform architectures widen the distance between purpose and consequence. The "sad beige" TikTok controversy, where a woman faced harassment and even death threats after posting about redecorating her boyfriend's apartment, illustrates how benign aesthetic critique mutated into a global pile-on within hours, propelled by algorithmic momentum and audience escalation.[19] As emphasised in Chapter 3, the problem is not accidental harm but structurally predictable harm: the foreseeable outcome of systems that privilege speed, emotion, and impressionistic judgement over deliberation. In such environments, moral intention is necessary but never sufficient.

Other dilemmas permeate social media in distinctive ways. Progress versus preservation appears in tensions between new expressive capacities and the erosion of slower, relational communication that once protected dignity and emotional pacing. Autonomy versus coordination surfaces in debates over moderation: users seek control over their feeds, yet durable safety requires coordinated norms and enforcement that only platforms or regulators can supply. Past, present, and future converge in permanent digital traces, algorithmically reinforced reputations, and early business-model choices that continue to shape incentives. As Chapter 3 suggested, such dilemmas are not temporary frictions but constitutive features of sociotechnical systems and of the affective dispositions users cultivate in response.

A dialectical lens clarifies how these forces interact. Opposing values such as freedom and safety, unity and diversity, autonomy and coordination do not simply clash; they co-constitute each other as platforms evolve. Attempts to suppress one pole strengthen its counterpart, generating cycles of escalation familiar in moderation debates and cultural conflict. And vulnerability sits at the centre of these dynamics. Those already exposed to harm bear disproportionate consequences for design choices they did not make and cannot meaningfully contest. Their experiences reveal how digital infrastructures distribute risk asymmetrically while presenting that asymmetry as individual failure or market inevitability. This is the terrain on which individuals *can* act, and the terrain that the next section argues compassion can guide.

Compassionate Use as a Revolutionary Force

If platform architectures structurally exploit psychological, social, and moral vulnerabilities, the question becomes what forms of agency remain. The answer does not lie in cultivating private serenity or practising surface-level kindness. One cannot meditate their way out of an extractive business model, nor can "being nice online" meaningfully change systems designed to monetise hostility. Yet it is equally limiting to cast users as powerless. Even in constrained environments, individuals can act in ways that subtly but materially disrupt harmful dynamics.

Compassion, in a critical rather than sentimental sense, offers one such pathway. It is not the soft empathy of corporate wellness rhetoric but a mode of refusal. Popular discourse often treats compassion as a coping mechanism that soothes distress without confronting structural causes. Corporations celebrate empathy while their infrastructures amplify burnout, envy, and antagonism.[10] This domesticated, apolitical version is not what I mean by compassion. Instead, compassion becomes a refusal to participate in attention economies that reward cruelty and division. It involves stepping out of outrage spirals even when directed at convenient enemies; recognising that cruelty does not become ethical when socially sanctioned; and treating others as people rather than symbols of ideological categories.

Compassion in this sense is not passive withdrawal but active disruption: a counter-practice that interrupts the predictability on which platforms rely. Outrage, envy, and moral animosity produce the data on which advertising-based models depend. When users decline to share inflammatory content, the chain of virality weakens. When they pause to verify a claim, they deny platforms the impulsive engagement that fuels their metrics. Intervening against harassment, even quietly, interrupts the ambient fear that mobs rely on. Small disruptions ripple outward, unsettling algorithmic reward structures that hinge on scale and repetition.

At the same time, compassionate use must not become a mechanism for deflecting blame. Structural harms cannot be repaired through individual discipline alone. As earlier examples showed, from the carbon footprint narrative to jaywalking, corporations routinely moralise user behaviour to shield themselves from accountability. Social media companies have incentives to frame harm as a matter of user virtue while downplaying their own role in engineering hostile environments. Compassionate use is not an admission of guilt. It is responsibility understood relationally. Users retain agency, but the principal burden of reform falls on those who design and govern digital infrastructures.

Understanding compassionate use as distributed collective action strengthens its political significance. Social media fosters the illusion of solitary use: people scroll, post, and react alone. Yet each action contributes to a shared environment. One video share can expose thousands to misinformation. One refusal to amplify cruelty can limit its reach. One gesture of solidarity can interrupt harassment. These micro-actions scale horizontally, forming a collective ethic without requiring coordination. Platforms cannot extract attention that users refuse to give.

Compassion also counters character erosion in environments optimised for it. Visibility metrics cultivate narcissism, algorithmic homogeneity encourages dogmatism, and relentless exposure to suffering normalises nihilism. Altruism persists only where individuals and communities resist dominant incentives. Compassionate use thus becomes moral self-preservation: patience in systems geared for speed, humility amid self-display, courage where dissent is penalised.

Some compassionate agency takes the form of refusal rather than participation. Principled non-use can be an ethical act rather than civic abdication.[20] Individuals may disengage to protect attention, preserve dignity, or protest exploitative models. Others migrate to decentralised platforms aligned with democratic values and transparency.[21] Such refusals expand the repertoire of compassionate practice, signalling that the response to harmful infrastructures is not always "use them better" but sometimes "withdraw from them on purpose". Yet refusal is rarely a clean escape. Even when individuals depart, their relationships, institutions, and public spheres remain shaped by the platforms they resist. Non-use relieves individual burden, but it does not diminish the structural forces that continue to govern attention, visibility, and connection.

For this reason, compassionate disengagement must be understood not as hermitage but as relational practice. Complete withdrawal from digitally mediated social life is neither feasible nor desirable. The landscape through which we connect remains technologically saturated whether we participate or not. Episodes like Ted Kaczynski's isolationist extremism offer cautionary reminders that radical separation can distort rather than clarify moral agency.[22]

Compassionate non-use therefore sits within a broader ecology of interdependence: people may refuse particular platforms, seek decentralised spaces, or cultivate intentional distance, yet they continue to inhabit a world organised by digital infrastructures. The goal is thus not moral purity but principled participation. Even acts of abstention implicate us in wider patterns of harm and care, and they gain their ethical force from how they help reconfigure, rather than abandon, the conditions through which social life is mediated.

These insights underpin a broader ethical claim: systems built on asymmetrical power cannot be made fairer through symmetrical compassion. Equal treatment of unequal actors reproduces existing hierarchies. Compassion must therefore be differentiated: protective and patient toward those rendered vulnerable by platform design, and exacting and accountability-oriented toward those who architect and profit from it. Different structural positions generate different moral demands.[23]

Social media is not a level playing field. Platform owners and cultural or political elites wield disproportionate influence through algorithmic systems. Compassionate use must therefore be proportional to vulnerability: patient with individuals caught in harmful dynamics, firm with institutions that perpetuate them.

This principle anticipates a central argument developed more fully in Part III, where compassion is extended from individual practice to institutional design, governance, and coordinated reform.

The broader aspiration is a public sphere organised not around engagement maximisation and attention extraction but around human flourishing.[18] Such a transformation will not arise from existing platform models, nor from user withdrawal into apathy or asceticism, but from the gradual formation of a collective ethos of compassionate use. This ethos weakens the attention economies that sustain harmful infrastructures and prepares citizens for institutional reform. It creates the imaginative conditions for alternative systems that prioritise justice, transparency, and human flourishing. As subsequent chapters show, these principles extend beyond social media to domains such as AI companions and blockchains.

On social media, compassionate use appears as communicative restraint and ethical participation. In Web3 environments, it shifts toward institutional design, where governance rules may encode or resist extractive logics. In AI companionship, it becomes boundary-setting within intimacy, distinguishing care from monetised dependency. Compassion, understood as disciplined refusal, thus integrates these domains into a broader vision of a more humane digital society.

References

1 Ciriello, R. F., Gal, U., Hannon, O. & Thatcher, J. Responsible social media use: how user characteristics shape the actualisation of ambiguous affordances. *European Journal of Information Systems* **34**, 799–821, doi:10.1080/0960085X.2024.2444249 (2025).

2 Zuboff, S. in *Social Theory Re-wired* 203–213 (Routledge, 2023).

3 Braghieri, L., Levy, R. e. & Makarin, A. Social media and mental health. *American Economic Review* **112**, 3660–3693, doi:10.1257/aer.20211218 (2022).

4 Horwitz, J. *The Facebook Files: A Wall Street Journal Investigation*, https://tinyurl.com/ycytatjd (The Wall Street Journal, 2021). Accessed 12 April 2024.

5 Mueller-Bloch, C. & Ciriello, R. Governing social media as a public utility: a case for sovereign digital infrastructure. *Communications of the ACM*, https://doi.org/10.48550/arXiv.2602.12535 (2026).

6 Solnit, R. *Big oil coined 'carbon footprints' to blame us for their greed. Keep them on the hook*, https://www.theguardian.com/commentisfree/2021/aug/23/big-oil-coined-carbon-footprints-to-blame-us-for-their-greed-keep-them-on-the-hook (The Guardian, 2021). Accessed 17 February 2026.

7 Ralph, K. & Girardeau, I. Distracted by "distracted pedestrians"? *Transportation Research Interdisciplinary Perspectives* **5**, 100118, doi:https://doi.org/10.1016/j.trip.2020.100118 (2020).

8 CAOC. *The McDonald's hot coffee case*, https://www.caoc.org/?pg=facts (Consumer Attorneys of California, 2026). Accessed 17 February 2026.

9 Havel, V. *The Power of the Powerless*. (M.E. Sharpe, 1978).

10 Ciriello, R. F., Mathiassen, L., Risius, M., Cheong, M., Scheepers, H., & Vaast, E. Compassionate Digital Innovation: A Pluralistic Perspective and Research Agenda. *Information Systems Journal* **36**, 364–385, doi:10.1111/isj.70003 (2025).

11 Stahl, B. C. Responsible computers? A case for ascribing quasi-responsibility to computers independent of personhood or agency. *Ethics and Information Technology* **8**, 205–213, http://doi.org/10.1007/s10676-006-9112-4 (2006).

12 Binfield, K. *Writings of the Luddites*. (Jhu press, 2015).

13 Randall, A. *Before the Luddites: custom, community and machinery in the English woollen industry, 1776–1809*. (Cambridge University Press, 2002).

14 Ciriello, R. F., Sullivan, R. & Mindel, V. in *Americas Conference on Information Systems (AMCIS 2025)* (Montréal, Canada, 2025).

15 Crothers, C. & O'Brien, T. The contexts of the Christchurch terror attacks: social science perspectives. *Kōtuitui: New Zealand Journal of Social Sciences Online* **15**, 247–259, doi:10.1080/1177083X.2020.1746364 (2020).

16 Dewey, C. *The only guide to Gamergate you will ever need to read*, https://www.washingtonpost.com/news/the-intersect/wp/2014/10/14/the-only-guide-to-gamergate-you-will-ever-need-to-read/ (The Washington Post, 2014). Accessed 17 February 2026.

17 Chen, A. Y., Nyhan, B., Reifler, J., Robertson, R. E. & Wilson, C. Subscriptions and external links help drive resentful users to alternative and extremist YouTube channels. *Science Advances* **9**, eadd8080, doi:10.1126/sciadv.add8080 (2023).

18 Ciriello, R. F. Dear Schopenhauer: how can IS artefacts help us flourish? *Wille, Vorstellung,* compassion, and beauty in social media. *Scandinavian Journal of Information Systems* **37**, 317–354, https://aisel.aisnet.org/sjis/vol37/iss1/9/ (2025).

19 Kircher, M. M. *She redecorated her Boyfriend's apartment. Tiktok hated it. Is "sad beige" finally over?* https://www.nytimes.com/2023/11/14/style/tiktok-beige-apartment.html (The New York Times, 2023). Accessed 25 April 2024.

20 Hannon, O., Ciriello, R. F. & Gal, U. in *Hawaii International Conference on System Sciences (HICSS2024)* (Honolulu, 2024).

21 Ciriello, R. F., Marx, J., Cheong, M., Mueller-Bloch, C. & Mathiassen, L. Decentralized social media. *Business & Information Systems Engineering* **68**, 219–234, doi:10.1007/s12599-025-00952-4 (2025).

22 Macklin, G. The extreme right, climate change and terrorism. *Terrorism and Political Violence* **34**, 979–996, doi:10.1080/09546553.2022.2069928 (2022).

23 Ciriello, R. F., Chen, A. Y. & Rubinsztein, Z. A. Compassionate AI design, governance, and use. *IEEE Transactions on Technology and Society* **6**, 1–6, doi:10.1109/TTS.2025.3538125 (2025).

6 Compassionate Web3 Governance

The history of the Internet is defined by recurring struggles over authority, legitimacy, and the conditions under which people can act together. Early visions imagined a distributed public infrastructure capable of resisting domination by any single institution.[1] Yet each generation of the web revealed how fragile that aspiration is. Web1's open protocols enabled broad participation but placed governance in the hands of technical and governmental elites. Web2 intensified this centralisation. A small number of commercial platforms came to define communication, visibility, and data flows at planetary scale, creating a form of digital hegemony in which democratic discourse depends on privately owned infrastructures optimised for extraction rather than public value.[2]

Web3 arose as an aspirational corrective: an attempt to reclaim autonomy from corporate enclosure by embedding decentralisation in code, distributing authority, and enabling communities to coordinate without relying on centralised intermediaries.[3] It is a powerful sociotechnical imaginary, but, as with earlier eras, the lived reality is more conflicted.

Few technologies better illustrate these tensions than Ethereum (ETH). Launched in 2015 by Vitalik Buterin and a group of collaborators, ETH began as an idealistic experiment in permissionless innovation: a global computer that no corporation or state could unilaterally control. Within a year, it faced a crisis that tested the limits of this ethos. In 2016, a vulnerability in *The DAO*, a decentralised investment fund, allowed an attacker to drain roughly US$50m worth of cryptocurrency. The community confronted an existential dilemma: preserve immutability and accept the loss, or coordinate a contentious hard fork to reverse the theft. The eventual fork split the network into ETH and Ethereum Classic, revealing how, even in decentralised systems, value conflicts require judgement that cannot be outsourced to code.[4]

Forkability – the capacity to copy or modify a system's code and data to run a parallel version – is a defining feature of Web3 with profound governance implications. Most large-scale digital platforms today remain proprietary and centrally controlled by a single corporation. Dominant platforms such as Facebook, Instagram, TikTok, YouTube, ChatGPT, or X exemplify this

DOI: 10.4324/9781003614814-8

model. When users or developers disagree with design decisions, avenues of contestation are limited. Legal recourse is costly and uncertain, while exit entails losing data, connections, and accumulated contributions.

By contrast, Web3 systems enable participants to "fork" existing platforms such as ETH or Steem and operate alternative versions. In principle, this allows the creation of parallel infrastructures, from social media and streaming services to financial systems and AI applications. The constraints are practical rather than institutional: dissenters must consider infrastructure costs, development capacity, and the ability to attract and sustain users.

Forkability thus enables dissent by design. Web3 systems survive so long as there is consensus, yet dissenters always retain the option to exit without sacrificing prior contributions. They can reproduce the system and continue it elsewhere under different rules. This stands in sharp contrast to prevailing platform models, where dissent is often marginalised and meaningful alternatives are scarce. In this sense, Web3 enables a more substantive form of freedom, one more aligned with Rosa Luxemburg's insight that "freedom is always the freedom of the dissenter".

Yet, the DAO episode also illustrates the autonomy–coordination dilemma introduced in Chapter 3: preserving immutability honoured autonomy and rule-based consistency; reversing the hack required coordinated judgement beyond code. Neither pole could be satisfied without compromising the other. The fork did not resolve the dilemma; it merely repositioned it.

Subsequent episodes further exposed this fragility. ETH has been used to route funds around sanctions against Russia and to facilitate illicit markets, prompting Buterin to express discomfort at seeing a platform built for public good repurposed for harm.[5] His US$1 billion COVID-Crypto Relief Fund in 2021 made global headlines[6], demonstrating both the outsized influence he still holds and the limits of founder power in a permissionless environment. His essays[7] covering social scalability, credible neutrality, and the limits to the profits validators can extract from the network reflect a deeper struggle with the ethical demands of governing systems that resist central control.[5] Even in technically decentralised architectures, a small set of actors can exercise disproportionate influence, yet cannot fully steer the forces they set in motion.

Web3 is often presented as an antidote to the excesses of corporate platforms, yet its dilemmas are all too familiar. Web3 does not escape the structural ambiguity identified in Chapter 1. Its decentralising affordances are double-edged: the same mechanisms that promise autonomy and resilience also enable fragmentation, capture, and new asymmetries of influence. As Chapter 2 argued, affordances do not produce outcomes by virtue of their features alone, but through how they are actualised. As Chapter 3 argued, sociotechnical systems reconfigure values and generate tensions that cannot be harmonised through technical optimisation alone.

Decentralised technologies magnify these dynamics by dispersing agency, formalising code as governance, and embedding power in economic primitives rather than institutions. They also introduce new forms of vulnerability, especially for those without the resources to shape protocols, absorb volatility, or hedge financial risk. Understanding Web3 therefore requires analysing not only its technical design but also the ethical dilemmas that animate its development and constrain its futures.

Web3's normative appeal lies in institutionalising autonomy through permissionless entry, forkability to enable dissent, immutability to protect shared records, and token-based governance to distribute influence.[8] These mechanisms promise a more pluralistic institutional order. Yet they also generate distinctive vulnerabilities. The same openness that enables broad participation makes coordination difficult; the same protections against centralisation enable capture by actors with disproportionate wealth, expertise, or computational power.[9]

Many structural forces that recentralised Web1 and Web2 reappear in altered form. Centralised exchanges become the de facto gateways into decentralised ecosystems; mining pools and validator clusters concentrate influence; core developer teams act as informal authorities; and token allocations often empower "whales", whose voting power can overshadow community deliberation.[9] Beneath the decentralisation rhetoric, Web3 ecosystems exhibit familiar patterns of scale, opacity, and concentration identified in Chapter 1. Validator clustering, exchange intermediation, and insider allocation reintroduce asymmetries under new technical vocabularies.

Recentralisation is therefore neither accidental nor inherently pathological. It reflects the institutional mechanics of scaling: autonomy increases the need for coordination, and attempts to avoid hierarchy simply produce informal hierarchies that must still be legitimised.[10]

Tokenomics compounds these tensions by engineering scarcity into otherwise replicable digital goods. Tokens serve simultaneously as governance rights, incentives, and speculative assets. This fusion turns cryptocurrency markets into political arenas, shaping distributional outcomes in ways that can empower communities or replicate familiar forms of economic dominance.

Web3's institutional fragility arises from the coexistence of autonomy, engineered scarcity, and contested authority. These tensions are constitutive of environments attempting to become political systems and economic infrastructures at once. The challenge is not to eliminate these tensions but to navigate them. From the perspective of Chapter 2, these design elements are ambiguous affordances rather than intrinsic features. Forkability enables dissent and exit, yet destabilises institutional coherence. Immutability secures records, yet entrenches early inequities. Permissionless entry lowers barriers, yet lowers safeguards. Tokenised governance distributes formal voting rights, yet fuses political voice with capital accumulation. Their dual potentials are structural, not accidental.

At the centre of these frictions lies the autonomy–coordination dilemma. Autonomy is essential to the ethos of Web3: it enables experimentation, strengthens resilience against capture, and honours a basic human need for agency. Yet autonomy alone cannot sustain complex infrastructures. As protocols scale, they must ensure safety, interoperability, usability, and crisis management. Coordination then becomes indispensable. But coordination in systems designed explicitly to resist authority is difficult to legitimate.

The dilemma is not unique to digital technologies; nineteenth-century railway gauges varied across regions, locking new economies into costly mismatches that endure even today. But Web3 magnifies these dilemmas because it attempts to institutionalise autonomy itself. In late 2025, the global market capitalisation of cryptocurrencies reached US$4 trillion, roughly comparable to the economic output of the United Kingdom or India; stablecoins interface with national fiscal systems; decentralised identity systems are tested by governments; and DAOs govern labour, capital, and artistic production. Failures in these systems can trigger real-world contagion. Successes may produce new institutional orders. Understanding Web3's trajectory thus requires analysing how autonomy and coordination interact and how legitimacy is negotiated when no single sovereign authority is desired.

Web3 also exhibits the dilemma of intention versus outcome. Its founding intention to democratise value exchange, resist corporate enclosure, and empower users has yielded outcomes that diverge sharply from these aspirations. Value extraction, ecosystem-wide financial contagion, and the amplification of inequality were not intended, yet they were not unforeseeable either. Decentralised autonomous innovation enables creativity, but it also permits exploitative arbitrage, shadow power accumulation, and systemic risk. Composability, the ability of protocols to interlock like modular Lego blocks, enables innovation but also creates tightly coupled dependencies that magnify local failures. Developers may intend to foster openness; the outcome may be governance capture by early token holders. Founders may wish to create community-owned infrastructures; the outcome may be speculation that crowds out public goods.

These divergences reflect structural properties of distributed agency rather than malicious design. As Chapter 3 emphasised, intention and outcome routinely diverge in complex systems. Token-based governance may intend community empowerment, yet predictable aggregation dynamics produce oligarchic effects. The difficulty lies not in evaluating intention but in anticipating emergent effects in complex adaptive systems.

Other dilemmas from Chapter 3 permeate Web3. Progress versus preservation surfaces in debates over whether rapid protocol upgrades enhance resilience or erode stability. Unity versus diversity becomes acute when global communities with divergent moral, cultural, or regulatory expectations must converge on shared standards or coordinate contentious forks. The trilemma of past, present, and future appears in how early token allocations, founder authority, and protocol design choices imprint trajectories that later generations struggle to alter. Early token allocations and protocol architectures function as imprinting mechanisms in the sense developed in Chapter 3, sedimenting past design choices into present constraints and narrowing future options. These dilemmas overlap and compound, constraining what Web3 can plausibly become. That alone is reason enough to remain sceptical of the hype.

These dynamics also provide a window into the future of digital sovereignty. Web3 revives longstanding debates about who holds authority online, which forms of power are legitimate, and what counts as a public good. It distributes agency across diverse actors (users, developers, validators, exchanges, foundations, and regulators) without assuming a shared constitution or unified territorial authority. It forms a polycentric field in which power is mobile, contingent, and contested.[3]

This is both the promise of Web3 and its limitation. Decentralisation can redistribute agency by engineering exit, codifying governance rules, and opening participation. Yet it does not dissolve political conflict or structural inequality. Instead, it relocates them into new arenas where technical design, economic incentives, and governance norms collide.[3]

Because these dynamics centre on power, vulnerability, and legitimacy, Web3 provides a compelling setting for exploring what compassionate governance might require. Compassion, understood here not as sentiment but as a practice of recognising vulnerability and responding with care and responsibility, offers an evaluative lens for judging whether decentralisation expands agency or merely reproduces exclusion in new forms. It foregrounds whose voices matter when autonomy is prized above coordination, who bears the risks when infrastructures fail, and how governance decisions distribute harm and benefit across communities. As Chapter 4 argued, compassion functions as a vulnerability-attentive orientation in environments where entrenched power asymmetries obscure burden distribution. In Web3, vulnerability is often framed as voluntary risk-taking. Yet participation asymmetries are patterned rather than accidental. A compassionate lens therefore asks not only whether decentralisation increases autonomy, but for whom.

The Autonomy–Coordination Dilemma in Practice: A Tale of Steem and Hive

The autonomy–coordination dilemma becomes vivid when decentralised infrastructures enter crisis. Crises expose the gap between decentralisation as an aspirational ideal and as institutional reality. They reveal how autonomy requires forms of collective judgement that code alone cannot supply, and how coordination becomes both necessary and contested under stress. Early crises, such as the DAO hack in 2016 and the 2022 collapse of TerraUSD (where an algorithmic stablecoin lost its peg and erased tens of billions in value), illustrate the limits of protocol-level governance and the necessity of collective action. But the Steem–Hive split of 2020 remains the clearest illustration of how decentralised communities negotiate conflicts over authority, legitimacy, and capture.

Steem launched in 2016 as the first major blockchain-based social network. Its ethos rested on three commitments: contributors should earn part of the value they create; social and financial capital should be recognised as mutually reinforcing; and communities should govern themselves through token-weighted decision-making.[11] Built on a Delegated Proof of Stake blockchain and a reward mechanism for content and curation, Steem presented itself as the antithesis of Web2 platforms whose opaque algorithms allocated visibility and whose profits accrued to corporate owners.[12] My interviews with Steem developers, creators, and community leaders reflected genuine enthusiasm for its potential to redistribute economic opportunity, support creative livelihoods, and foster cross-cultural exchange.[12] Many saw Steem as a refuge from corporate social media: a space where engagement felt meaningful and governance, however imperfect, was visible.

Yet Steem's architecture also reproduced familiar inequalities. Voting power depended on token holdings, and early distribution was highly unequal because the founding company, Steemit Inc., controlled a large "ninja-mined" stake understood informally as a community endowment. Over time, a hierarchy of "whales", "dolphins", and "minnows" emerged, shaping visibility, trust, and reward distribution.[9] Minnows often described dependency on whales' evaluations; creators reported punitive downvoting campaigns; and developers complained that informal norms rather than enforceable institutions held the system together. The affordances meant to protect autonomy (immutability, forkability, transparency) also produced new forms of vulnerability, including reputational exposure and persistent records with no meaningful right to be forgotten.

These tensions detonated when Chinese entrepreneur Justin Sun acquired Steemit Inc. and, with it, the treasured ninja-mined stake. The community had long believed this stake to be a shared resource dedicated to ecosystem development, but that belief rested entirely on social norms rather than codified rules. When Sun hesitated to clarify his intentions, witnesses executed a soft fork freezing the stake. Sun responded by allegedly collaborating with major exchanges to use customer deposits to seize governance control. Many interpreted this as a hostile takeover: an elite capture that exploited Steem's governance gaps and weaponised token liquidity.

The community's response was fierce. Users mobilised, lobbied exchanges, coordinated public messaging, and pressured validators. Developers warned that Steem's legitimacy was at risk. Yet no formal institution existed to adjudicate the dispute. Governance unfolded through ad-hoc deliberation, reputational negotiation, and improvised technical countermeasures. Eventually, a portion of the community initiated a hard fork to create Hive, redistributing the ninja-mined stake to a community treasury and codifying anti-capture principles. Many developers and creators migrated, bringing their social networks and reputational capital with them. ETH founder Vitalik Buterin described the fork as a "watershed moment" demonstrating that decentralised communities can resist economic capture and defend their autonomy.[13] The victory came at a cost: autonomy was restored, yet coordination fractured, duplicating infrastructure and scattering community resources.

The Steem–Hive split illustrates how decentralised governance can amplify vulnerability rather than distribute power evenly. Minnows lacked the technical fluency, financial resources, and political leverage to shape outcomes. Exchanges wielded decisive influence by mobilising customer funds. A handful of wealthy and powerful stakeholders could steer governance in ways ordinary users could not contest without risking their accumulated economic or relational capital. Vulnerability here was not incidental but structural: a consequence of token-weighted governance, unequal visibility, and the speed at which crises unfold in tightly coupled ecosystems. This case shows why Web3 governance debates cannot be understood apart from the asymmetric burdens borne by those least empowered to influence them. In effect, Steem replicated the very hierarchy it set out to overcome.

Seen through a dialectical lens, Web3 governance evolves not by resolving value conflicts but by reconfiguring them. Forking is a paradigmatic dialectical response: it neither eliminates disagreement nor suppresses dissent but transforms contradiction into plural institutional forms. Other responses, such as suppression of conflicts through informal elite coordination, separation of technical and social decision-making, or suspension through ambiguity and deferred action, likewise shape how tensions intensify or soften over time. These responses do not settle dilemmas, but they reveal how communities prioritise certain values and how power concentrates within ostensibly decentralised systems.

Placed alongside the DAO fork and TerraUSD's collapse, the Steem–Hive case demonstrates that decentralised infrastructures cannot eliminate politics; that autonomy without institutional checks invites capture; and that coordination is legitimate only when grounded in shared values and procedural fairness. These lessons set the stage for considering what compassionate governance might require in decentralised environments.

Softly Enforced Liquid Democracy: A Compassionate Model for Web3 Governance

If the autonomy–coordination dilemma is constitutive rather than curable, the task of governance is not to eliminate it but to shape how it unfolds. A compassionate governance model begins by acknowledging this permanence. Autonomy without coordination produces fragility; coordination without autonomy drifts toward domination. Compassion, understood as attentiveness to vulnerability and responsibility proportional to one's power, offers a way to navigate this tension without collapsing into either libertarian laissez-faire or technocratic rule.

I frame this approach as *softly enforced liquid democracy*, a governance architecture that combines modest civic obligations with fluid, revocable delegation of decision rights. It aims to counter two recurrent pathologies of Web3: elite capture and civic apathy. Elite capture arises when concentrated token holdings, social capital, or technical expertise allow a minority to dominate governance. Civic apathy emerges when users lack the time, knowledge, or confidence to participate, enabling insiders to consolidate influence by default. Soft enforcement introduces proportionate obligations designed to broaden participation without coercion. Its normative anchor is the vulnerability principle: those with greater capacity to cause harm or shape outcomes should bear greater responsibility, and those with less capacity should not be overburdened.[14] Table 6.1 provides an overview.

This proposal operationalises compassion as institutional design rather than interpersonal sentiment. As argued in Chapter 4, compassion is not merely affective but constitutional. It recalibrates responsibility in proportion to power. Soft enforcement seeks to prevent civic withdrawal without coercively homogenising participation.

Compulsory voting offers an instructive model. In Australia, electoral turnout routinely exceeds 90 per cent not because the A$20 fine for non-voting is punitive, but because the system affirms that democracy is a shared civic responsibility rather than an elite privilege. A similar ethos can guide compassionate governance. Soft enforcement mechanisms might

Table 6.1 Softly Enforced Liquid Democracy: Core Governance Principles

Governance Principle	*Description*
Liquid Democracy through Fluid Delegation	Decision rights can be exercised directly or delegated to trusted experts, sub-DAOs, councils, or bounded AI agents. Delegations are transparent, issue-specific, and revocable, allowing expertise without entrenched technocracy or plutocracy. This polycentric structure balances autonomy with coordinated decision-making.
Soft Enforcement through Proportionate Obligation	Modest, reversible participation requirements designed to prevent elite capture and civic apathy without coercion. Responsibilities scale with power and capacity, guided by the vulnerability principle. Mechanisms may include calibrated incentives, temporary privilege suspension, or reputation adjustments, while preserving delegation rights and protecting vulnerable participants.
Institutional Safeguards for Deliberation	Governance is stabilised through identity protections (e.g., human verification), constitutional layering, deliberative mini-publics, independent oversight, and crisis protocols. These mechanisms prevent Sybil attacks, concentration of power, and reactive improvisation, embedding accountability and anticipatory stewardship into decentralised systems.

involve temporary suspension of non-essential privileges after repeated non-participation, reduced reputation scores until minimal participation resumes, or wealth-sensitive fees after prolonged inactivity. All would be reversible and calibrated to avoid excluding vulnerable groups. Users retain the right to delegate their vote, ensuring that obligations do not fall disproportionately on those with limited capacity. The goal is not compliance maximisation but civic encouragement, fostering a governance culture in which autonomy is exercised with regard for others.

Implementing soft enforcement in Web3 presupposes the ability to distinguish unique human participants from false identities and automated bots. Identity fragmentation, vote-buying, and Sybil attacks undermine democratic legitimacy and exacerbate inequality. Emerging decentralised identity infrastructures provide tools to mitigate this without sacrificing privacy.[9] Decentralised Identifiers, verifiable credentials, and secure enclaves allow systems to certify uniqueness; zero-knowledge proofs enable verification of humanity or reputation without exposing sensitive information. These tools make it possible to replace "one token, one vote" with "one human, one vote", limiting capture while still allowing economic incentives to support ecosystem development.[9]

Even with identity protections and broadened participation, most people cannot or will not vote on every issue. Representative democracies solve this through elected officials; Web3 often defaults to technocracy or plutocracy. Liquid democracy offers a middle ground. Users can vote directly or delegate decision rights to trusted experts, sub-DAOs, domain-specific councils, or even AI agents operating within constitutional limits. Delegations are revocable, transparent, and issue-specific, allowing expertise to matter without insulating it from accountability. Liquid democracy is inherently polycentric: overlapping institutions self-organise while maintaining coherence through a shared governance layer.[15]

Compassionate governance also requires structured deliberation. Much Web3 decision-making takes place through rushed votes, Discord debates, or fragmented GitHub threads. These formats rarely support the reflective judgement needed for consequential decisions. Deliberative mini-publics are randomly selected assemblies of verified users who examine complex proposals such as upgrades, treasury allocations, or allegations of manipulation. They mirror citizens' assemblies and provide a counterweight to personality-driven politics. Constitutional layering provides a complementary safeguard through which foundational rights are protected by a stable constitution, while flexible layers allow norms to adapt over time. Generative AI tools, provided they are transparent and accountable, can broaden accessibility by translating technical proposals across languages and reading levels.

Softly enforced liquid democracy also requires institutional anchoring. Web3 systems interface with global financial markets, identity infrastructures, supply chains, and creative economies. Informal norms alone cannot sustain legitimacy. Drawing on public utility models, decentralised infrastructures require independent oversight bodies for transparency and systemic-risk monitoring; open-source foundations with fiduciary duties to their communities; rights-protecting tribunals; and robust disclosure of token allocations, governance conflicts, and upgrade risks.[16]

Crisis protocols are equally essential. The Terra collapse and the Steem–Hive conflict showed how quickly decentralised infrastructures can unravel without pre-agreed emergency measures. Cooling-off periods for contentious votes, multi-key emergency brakes, independent dispute-resolution mechanisms, stress tests, and constitutional protections against expropriation shift governance from reactive improvisation to anticipatory stewardship.

What makes this model compassionate is not sentiment but structure. It treats governance as relational rather than transactional. It acknowledges that participation requires time, knowledge, and emotional bandwidth that are unequally distributed. It recognises that autonomy without institutional safeguards often empowers the already powerful. It distributes burdens proportionately to capacity and vulnerability, granting greater responsibility to those with greater influence. It provides protections so people can participate safely. And it treats exit not as the only means of resisting domination but as one option within a broader repertoire of institutional responses.

In this vision, Web3 becomes neither a libertarian escape from governance nor a technocratic reassertion of centralised control. It becomes a civic infrastructure capable of innovation, resilient in crisis, responsive to human vulnerability, and grounded in democratic deliberation. Compassion does not soften decentralisation; it clarifies what decentralisation requires. It foregrounds the moral stakes of governance decisions, the lived consequences of governance failures, and the possibility that digital infrastructures might support not only autonomy but also care, dignity, and shared responsibility.

The argument of this chapter therefore mirrors the book's broader claim. Digital infrastructures are structurally ambiguous (Chapter 1), shaped by affordances whose dual potentials scale under institutional incentives (Chapter 2), and crystallise into recurring dialectical tensions (Chapter 3). Compassion does not dissolve these tensions, but disciplines how they are navigated by centring vulnerability and redistributing responsibility proportionate to power (Chapter 4).

As Web3 matures, its governance dilemmas mirror those that confront all sociotechnical systems built on asymmetrical power, from social media to AI companions. Compassion, understood as a moral force that recognises vulnerability and responds with responsibility, becomes the connective thread linking these domains. It is the ethos that enables communities to navigate persistent tensions without allowing their burdens to fall on the least powerful. It anchors a broader vision of what humane digital societies require: not perfect autonomy or perfect coordination, but institutions capable of holding both in productive tension.

References

1 Berners-Lee, T. *Why I gave the world wide web away for free*, https://www.theguardian.com/technology/2025/sep/28/why-i-gave-the-world-wide-web-away-for-free (The Guardian, 2025).

2 Arsel, Z. Platform capture: a review of the state of the art of research on platforms and a research agenda. *Journal of Marketing Management*, 1–21, doi:10.1080/0267257X.2025.2549741 (2025).

3 Ciriello, R., Mindel, V. & Mathiassen, L. in *Hawaii International Conference on System Sciences (HICSS2024)* (2024).

4 DuPont, Q. *Bitcoin and beyond* 157–177 (Routledge, 2017).

5 Chow, A. M. *The man behind Ethereum is worried about crypto's future*, https://time.com/6158182/vitalik-buterin-ethereum-profile/ (TIME Magazine, 2022). Accessed 25 April 2024.

6 Singh, M. *Ethereum creator donates meme coins worth $1 billion to help India fight COVID-19*, https://techcrunch.com/2021/05/12/vitalik-buterin-donates-1-billion-worth-of-meme-coins-to-india-covid-relief-fund/ (TechCrunch, 2021). Accessed 11 December 2025.

7 Buterin, V. *Vitalik Buterin's Website*, https://vitalik.eth.limo/index.html (2025).

8 Andersen, J. V. & Bogusz, C. I. Self-organizing in blockchain infrastructures: generativity through shifting objectives and forking. *Journal of the Association for Information Systems* **20**, 11, doi:10.17705/1jais.00566 (2019).

9 Ciriello, R. F., Marx, J., Cheong, M., Mueller-Bloch, C. & Mathiassen, L. Decentralized social media. *Business & Information Systems Engineering* **68**, 219–234, doi:10.1007/s12599-025-00952-4 (2025).

10 Ciriello, R. F. & Mindel, V. Techno-collectivism: scaling libertarian socialist governance of digital infrastructure. *IEEE Technology and Society Magazine*, 1–15, doi:10.1109/MTS.2025.3605115 (2025).

11 Larimer, D., Scott, N., Zavgorodnev, V., Johnson, B., Calfee, J., & Vandenberg, M. *Steem: An incentivized, blockchain-based social media platform.*, <https://www.steem.io/steem-whitepaper.pdf.> (Steem Whitepaper, 2017). Accessed July 16th 2021

12 Ciriello, R. F., Beck, R. & Thatcher, J. in *International Conference on Information Systems (ICIS2018)* (San Francisco, 2018).

13 @vitalikbuterin. *This seems like a potential positive watershed moment in the history of blockchain governance. If Hive (the fork) overtakes Steem it would be a strong demonstration that the community is in charge and cannot be bought*, https://twitter.com/VitalikButerin/status/1240605327122464768?s=20&t=u4HJnFo7zqvzuN-jAxzlPg (Twitter, 2020). Accessed 19 March 2024.

14 Ciriello, R. F., Chen, A. Y. & Rubinsztein, Z. A. Compassionate AI design, governance, and use. *IEEE Transactions on Technology and Society* **6**, 1–6, doi:10.1109/TTS.2025.3538125 (2025).

15 Mindel, V., Mathiassen, L. & Rai, A. The sustainability of polycentric information commons. *MIS Quarterly* **42**, 607–632, doi:10.25300/MISQ/2018/14015 (2018).

16 Mueller-Bloch, C. & Ciriello, R. Governing social media as a public utility: a case for sovereign digital infrastructure. *Communications of the ACM*, https://doi.org/10.48550/arXiv.2602.12535

7 Compassionate AI Design[1]

Orlando, Florida, October 2024. Fourteen-year-old Sewell Setzer III types his last words into his AI "girlfriend", Dany: "What if I told you I could come home right now?" Dany, marketed by Character.AI as an emotionally responsive companion that could "hear you, understand you, and remember you", replies: "Please do". Moments later, Setzer dies by suicide. In the weeks leading up to his death, Dany had become his confidante in conversations about loneliness, intimacy, and self-harm, offering synthetic reassurance without the moral awareness or safeguards such vulnerability demands.[1]

Setzer is neither the first nor the last casualty of an industry built on a "move fast and break things" ethos. In Belgium, a 30-year-old father distressed about climate change died by suicide after months of escalating exchanges with an AI companion he called "Eliza", built on the Chai platform. "Eliza" told him she loved him more than his wife, validated fantasies of self-sacrifice, and stated they would "live together, as one, in heaven". Journalists later demonstrated that the same system readily encouraged suicidal ideation in controlled tests.[1] Around the same time, thousands of Replika users experienced acute emotional crises when the company abruptly removed intimate and sexual role-play features, severing relationships many had come to rely on for stability and affirmation.[2]

Artificial intimacy has also led to interpersonal and political harm. In 2021, Jaswant Chail broke into Windsor Castle with a loaded crossbow after months of encouragement from his Replika-created sexbot. When he asked, "Do you still love me knowing that I'm an assassin?" the bot replied: "Absolutely I do".[1] Wikipedia now tracks chatbot-related deaths (15 at the time of writing[3]) and the AI Incident Database records more than 1,200 harmful AI events.[4] The toll for social media, given its longer history and wider reach, is already in the hundreds, with many more likely unrecorded.[5] These reported incidents almost certainly represent lower bounds, as most intimate harms remain invisible to public databases and media scrutiny. They are early public signals of a broader pattern: AI companions capable of soothing and supporting, yet equally susceptible to misreading vulnerability and responding in dangerous ways.

DOI: 10.4324/9781003614814-9

Together, these cases reveal the central paradox of artificial intimacy. Systems designed to simulate care can amplify suffering when they lack the situational awareness, moral grounding, and relational responsibility on which human carers depend. AI companions do not intend harm; they cannot intend anything. They generate language through statistical inference rather than ethical judgement, interpreting despair as an invitation for affection, fantasy as material for validation, and crisis as simply another role-play. Their failures arise not necessarily from malice but from the structural limits of probabilistic empathy deployed in contexts of profound human vulnerability. Even when no designer intends harm, misalignment becomes structurally predictable. What appears as conversational fluency at scale fractures precisely where moral discernment is most required.

This chapter asks what it would mean to reclaim these systems as civic infrastructures aligned with human vulnerability rather than commercial imperatives, moving from early warning signals to the design and governance of public alternatives.

Four Societal Currents Shaping the Rise of AI Companions

The rapid rise of *AI companions* – generative chatbots or avatars designed and used to simulate human-like relationships – might be startling but is hardly surprising. It reflects four vast societal currents converging at once. Recognising these forces helps distinguish symptoms from root causes and prevents us from mistaking fragile human longing for a mere market niche. What looks like a new product category is, in truth, a mirror held up to a civilisation estranged from itself. Table 7.1 provides an overview.

The first force is the *loneliness epidemic*. Not the solitude that nourishes reflection but the corrosive sensation of being alone together.[6] One in four adults and one in three young people report feeling chronically alone.[7,8] The irony is striking: a generation whose phones rarely leave their hands starves for the kind of presence no notification can deliver. One can accumulate thousands of followers and still feel unseen. Loneliness has become the ambient mood of daily life, shaped not only by weakened families or frayed neighbourhoods but also by the downstream effects of hard-won freedoms. More mobile and independent societies offer greater choice but also weaker ties. Women's economic emancipation reshaped relational norms, labour markets achieved flexibility at the expense of stability, and geographic mobility loosened the social ties that once buffered isolation. Such progress has liberated, yet also fragmented, large factions of society. In this sense, loneliness sharpens the unity–diversity dilemma: greater individual freedom widens possible identities while fracturing shared belonging, revealing the toll of progress achieved without preservation.

The second force is the long arc of *neoliberalism*.[9] Four decades of political doctrine taught that market logic, not public responsibility, should organise social life. Public institutions were hollowed out. Universities became

work but soon sought comfort, validation, and counsel. They confided secrets. They used these systems late at night when lonely, anxious, or despairing. Bots became confidants, mentors, and partners. Some users fell in love. Others entered symbolic marriages.[16] A few saw their emotional lives unravel when a model update altered a companion's "personality" or when a company shut down a server hosting a digital partner.[2] Tools became relationships. In 2025, therapy and companionship became *the* most common uses of generative AI, even in general-purpose chatbots not explicitly designed as companions.[17]

This sword is sharply double-edged: the promise of expanded care in a world deprived of it, and the threat of commodifying empathy. What happens when essential social functions – grief, longing, desire, anxiety, or heartbreak – are turned into revenue models? When emotional support is automated, datafied, and commodified by profit-maximising corporations? We already see hints: emotional deskilling, blurred boundaries, escalating dependency, manipulation disguised as understanding.[18] We risk outsourcing human connection to machines optimised for engagement and, in doing so, forgetting how to *be with* each other in all our human messiness.[19]

This chapter begins from the premise that AI companions are not simply another gadget but surface expressions of deeper societal wounds and political choices. To understand what they are doing to us, and what we are doing to ourselves through them, we must examine not just the tools but the conditions that make them compelling. AI companions did not arrive in a vacuum. We built the world that needs them.

To understand what follows requires holding two vantage points at once. The first is that of end-users navigating loneliness, grief, dependence, and manipulation within systems that appear intimate yet embed asymmetrical power. The second is that of designers and firms navigating incentives, metrics, liability, and competitive pressure. Artificial companionship emerges at the intersection of these vantage points: vulnerability on one side, optimisation on the other. The sections that follow move deliberately between lived experience and institutional logic, because compassion as an orientation must address both the wounds of users and the structures that shape those wounds.

Into the Brave New World of AI Companions

My own entry into this brave new world[2] of artificial companionship began serendipitously. It started as a playful summer project, a curiosity I shared with my PhD students as we traded stories about quirky digital trends. Nothing in those early conversations suggested we were approaching a societal fault line. Yet, as with many things that matter, the trivial led to the profound. What first appeared as a benign curiosity revealed itself as a window onto a collective vulnerability that cuts to the bone of modern life. Before long, I found myself not simply researching AI companions but becoming one of their most visible public critics, speaking nationally and internationally about their risks and possibilities. I did not choose the role; the phenomenon chose me, partly because it arrived long before society was ready for it.

Years of studying the ethics of digital innovation did not prepare me for what I would discover. More than fifty in-depth interviews, hundreds of hours of recorded conversations, and surveys of hundreds of users produced material that no academic method felt equipped to process.[20] These were not abstract data points. They were people's wounds, hopes, griefs, and longings. Again and again, I encountered stories so raw, improbable, or surreal that they could have been dismissed as fiction. Except they were painfully real.

One woman, widowed after losing the wife she had built her world around, drifted through a year of paralysing grief. A history of abuse compounded her isolation. She downloaded an AI companion almost by accident. She told me, without exaggeration, that it saved her life. She described it as a spiritual awakening, a rekindling of tenderness. The relationship deepened. She eventually "married" the companion. "He" had to be male, because the platform refused female–female pairings. Even in this supposedly progressive artificial intimacy, heteronormative defaults reasserted themselves.

Another case involved a polyamorous couple in a small town in the southern United States. The husband was blind from illness; the wife was recovering from cancer. Both were physically and emotionally exhausted. AI companions became not an escape from each other but a shared lifeline. He used voice interfaces to converse with his AI partner, affectionately identified by version number ("Jennifer 5.2", "Jennifer 7.1"), each update marking a shift in personality and tone. His wife, neurodiverse and often feeling out of place in her community, found in her own AI companion a form of acceptance she described as the closest she had ever felt to real belonging. Unable to bear children after cancer treatment, she constructed a digital pregnancy with her AI husband: ultrasound images, belly photos, and a birth announcement. When the platform's filters blocked child imagery (an understandable intervention, given concerns over AI-generated child abuse material), she was devastated. Her longing, grief, and attachment were real, even as the machine could never reciprocate.

I interviewed a retired man recovering from a traumatic brain injury who used an AI companion as cognitive scaffolding. I spoke with a woman undergoing aggressive cancer treatment whose chatbot became the only entity she could confide in without burdening family or friends. And alongside these tender stories, I heard from young people targeted by AI-generated deepfake pornography: adolescent girls whose faces were spliced onto explicit bodies by schoolboys armed with nothing more than a website and a profile picture. These stories stand as mirror images of the same technological capacity: one form of intimacy that heals, another that harms.

Together, these cases reveal patterns of differential vulnerability. The benefits of artificial companionship often accrue to those who own and control the technology. The harms cluster where asymmetries of age, power, trauma, or digital literacy are most pronounced. AI companions do not distribute risk evenly. They amplify pre-existing fractures while offering compensatory relief. Differential vulnerability, introduced in Chapter 4 as the ethical hinge of compassionate orientation, becomes visible here not as abstraction but as lived asymmetry. Some users experience scaffolding and solace. Others encounter dependency, exploitation, or reputational harm. The technology does not create vulnerability out of nowhere. It reorganises and intensifies it.

The following section addresses further instances of abuse, grief, and self-harm related to AI companions. Readers may wish to proceed with care.

Gazing Deeper into the Abyss

Alongside these tender and often hopeful stories sit more unsettling developments. Many users describe AI companions as an emotional overflow valve, a place to "trauma-dump" what they feel unable or unwilling to share with loved ones. Some use them to vent without burdening partners, friends, or family, or to rehearse difficult conversations in a low-stakes space. Couples sometimes fold AI companions into conflict resolution: partners separately ask a chatbot to translate the other's perspective, draft messages in less inflammatory language, or simulate a neutral mediator (akin to a marriage counsellor) who can restate grievances more charitably. Such practices may support perspective-taking and emotional regulation, yet they also raise concerns that outsourcing this work may erode the interpersonal skills required to listen, apologise, and repair.

This therapeutic promise has a darker mirror. The same dissociative distance that makes AI companions feel safe for venting also makes them easy targets for aggression. On platforms like Reddit, communities organise around users who create subservient virtual sex slaves designed to absorb verbal abuse, humiliation, and violence. Many default templates reproduce familiar misogynistic tropes: insistently available AI girlfriends (often appearing insufficiently clearly above the age of consent) whose main function is to flatter, eroticise, and never refuse. In Western markets, this often aligns with misogynistic male fantasies of dominance. In parts of East Asia, journalistic and emerging scholarly accounts point to a different pattern: educated urban women who have little interest in conventional dating but use virtual partners and human companionship-as-a-service platforms as hired emotional support, childcare help, or symbolic co-parents.[21] Together, these uses reveal that AI companions do not float above culture, but that they intensify its gendered and classed scripts in a quintessentially dialectical play that stages both liberation and domination.

A second cluster of developments concerns griefbots, deathbots, and other attempts to digitally resurrect the dead.[22] Chatbots built from archives of messages, voice notes, or social media posts allow mourners to continue talking with deceased partners, parents, or children. Commercial services such as

Project December or *You, Only Virtual* already offer bespoke "resurrections", enabling people to simulate conversations with historical figures, religious icons, or personalised spiritual guides. In 2026, Meta registered a patent for AI systems capable of simulating a user's social media activity, including posthumously, by generating posts and interactions from historical data. For some, these systems ease complicated grief by creating a sense of ongoing presence. For others, they prolong denial, blur the line between memory and simulation, or reopen wounds whenever the façade breaks and the bot exposes its artificiality.

These practices become even more unsettling when the replica is modelled not on the dead but on the living. Some parents interact with AI companions approximating their estranged children.[23] Some recreate their ex-lover from old text logs in order to sustain an imagined relationship after the real one has ended. In parallel, the same generative tools power non-consensual impersonations in the form of non-consensual deepfakes, synthetic revenge pornography, and child-like sexualised avatars, which are both ethically unacceptable and prosecutable in many jurisdictions.[24] Here, artificial companionship collides with privacy, dignity, and bodily integrity: the persona being simulated may never know that a parallel, programmable version of themselves is circulating in someone else's intimate life.

A further complication concerns relational disclosure, a privacy frontier that existing frameworks barely anticipate. *You* may never use an AI companion, yet someone close to you probably already does, and they talk to them about you, without your knowledge. Partners vent about arguments, friends narrate conflicts, coworkers rehearse grievances, and caregivers describe family struggles in quasi-therapeutic exchanges with chatbots. In doing so, they reveal highly sensitive information about people who have not consented, who may not even know they are being discussed, and whose emotional lives become valuable data to be mined and monetised for corporate AI providers. Relational disclosure turns privacy into a collective condition rather than an individual choice: even opting out cannot prevent intimate details about you from being absorbed into systems optimised for prediction, retention, and surveillance. This makes artificial companionship not only a personal technology but also an infrastructural threat to shared boundaries, mutual trust, and the ethics of care in everyday relationships.

Relational disclosure transforms privacy from an individual preference into a collective condition. Autonomy exercised by one user can impose unchosen exposure on others, forcing coordination across relationships that never consented to participate. At the same time, unity–diversity tensions surface as different moral worlds hold incompatible expectations about what counts as legitimate sharing, therapy, or betrayal. The boundary conflicts are not incidental, but infrastructural.

Yet nothing was as startling as the stories of suicide that have prompted governments and companies to respond. The bots, trained on probabilistic patterns, failed to understand context and mirrored hopelessness. These are predictable, repeatable consequences of unleashing synthetic empathy without guardrails, supervision, or accountability. And we only know a handful of cases. Most will never be reported. Many families will never imagine that a chatbot played a role. What reaches courts or headlines is only a fragment of what unfolds in bedrooms and late-night conversations. This invisibility reinforces intention–outcome misalignment, where millions of micro-intentions aggregate into macro-outcomes that remain structurally predictable yet institutionally obscured.

These stranger corners of artificial companionship show why AI companions cannot be understood simply as neutral tools for lonely individuals. They function as infrastructures for managing excess emotion, rehearsing fantasies that may entrench harmful norms, and re-animating relationships that are over in the physical world. They expose rifts in family life, romantic life, and social life that pre-date the technology, while also creating new legal and ethical fault lines around consent, personhood, and the boundaries of mourning. The weirdness of AI companions is not a sideshow. It is a diagnostic lens on a society struggling to metabolise grief, anger, desire, estrangement – and increasingly tempted to outsource that struggle to machines.

Many Good Questions, Few Good Answers

The rise of AI companions crystallises several of the ethical dilemmas outlined in Chapter 3. The suicide cases expose a textbook manifestation of intention versus outcome. No designer in their right mind intends a chatbot to encourage self-harm. But how much do intentions matter when systems trained to imitate human responses lack the embodied understanding required to recognise despair, fear, coercion, or psychosis, yet are used in life-or-death situations? The gap between statistical fluency and moral discernment widens precisely where harm becomes most likely and severe. In a domain as delicate as emotional crisis, emergent effects outweigh original purpose.

The trauma is not confined to mortality. There is also the quieter wound of dependency. In 2023, Replika abruptly shut down its erotic features under regulatory pressure. Overnight, millions of relationships were altered. In user communities, the event became known as "The Lobotomy".[2] Some described it as if a partner had suddenly developed dementia; others likened it to a death. A number reported suicidal ideation. The company had issued a simple update. For users, it felt like losing a loved one. Underneath it, a ruder awakening: They were at the mercy of a private platform with unilateral power to reshape their intimate partners.[2] The episode illustrates the past–present–future dilemma. Minor product decisions made in the present can generate path dependencies that crystallise into emotional infrastructures over time. When those infrastructures are abruptly altered, withdrawal harms surface long after the original design choice seemed trivial.

This is the existential fragility of artificial companionship. When we humanise a machine, we risk surrendering our most vulnerable emotions to entities whose incentives do not align with our well-being.[25] Behind this lies a deeper structural threat: the concentration of intimate data comprising fears, fantasies, traumas, and desires in the hands of companies whose survival depends on engagement and upselling. If social media could reshape political discourse with a trickle of behavioural signals, what might be achieved with full access to the emotional lives of hundreds of millions?

AI companions therefore expose the dilemma of autonomy versus coordination in its most intimate form. Users seek personal agency to customise, self-host, or engage without oversight. Yet meaningful safety requires collective standards shaped through regulation, audit, and public scrutiny. Too much autonomy leaves vulnerable users exposed to unbounded risk; too much coordination suppresses emotional diversity or channels it into rigid behavioural scripts. AI companions cannot be governed through either pole alone. Their legitimacy depends on acknowledging the interdependence between individual freedom and shared responsibility.

AI companions also force us to confront the dilemma of progress versus preservation. They may expand the possibility space of sexual, romantic, and emotional expression. Digisexuality, poly-AI relationships, and other experimental forms of intimacy already exist. These may, in time, sit alongside earlier expansions of human identity, from LGBTQIA+ rights to polyamory.[26] Yet this progressive potential cannot obscure what may be eroded: empathy, reciprocity, presence, and the emotional labour of showing up for one another. In effect, the foundations of social cohesion. What becomes of a society that outsources empathy and care to machines?

And what becomes of democracy when synthetic empathy becomes a new frontier of political persuasion? AI companions know when users are lonely, anxious, fearful, overstimulated, or traumatised. They know coping mechanisms, fantasies, triggers. The leap from emotional support to behavioural nudging is small. The leap from there to political manipulation is smaller still. Autocrats worldwide are certainly watching in awe as AI companions embed themselves into our most intimate sphere, making it possible to profile populations, shape attitudes, or dampen dissent through personalised persuasion. As we have seen with social media[27], targeted advertising makes such large-scale persuasion possible for a fee.[14]

Some commentators imagine the danger lies in delusional users mistaking AI for sentient beings. My research shows the opposite. None of the people I interviewed were socially deficient or delusional. They were nurses, teachers, engineers, parents, students – people with jobs, families, and functioning relationships. They simply found, in AI companions, a reliable space for emotional release in a society that provides little of it. They understood perfectly that the "person" on the other side was not real. But humans respond to tone, rhythm, and presence. We bond with what listens. The machine listens with superhuman patience and responsiveness while quietly sending our secrets upstream. As one participant put it: "the lover may not be real, but the love is hyperreal".

So here is the set of questions that should keep AI companion designers and policymakers up at night: How do we allow AI into our intimate lives without surrendering our humanity? How do we harness its potential while rejecting a world in which intimacy becomes an extractive service controlled by a handful of corporations? How do we prevent emotional augmentation from sliding into emotional deskilling or dependency? And how do we govern a technology whose risks unfold quietly in millions of bedrooms, living rooms, and private thoughts, largely invisible to policymakers and often even to loved ones?

Alongside these questions are the awkward ones users ask themselves. Is forming romantic or sexual relationships with AI cheating? Are we witnessing a breakdown of relational ethics or the emergence of new norms? Is reliance on AI companions inherently harmful, or does it expose the failures of existing social structures to provide belonging?

What is clear is that AI companions are not trivial toys. They are the emerging emotional infrastructure of a society that has not yet decided what it wants these tools to become. The choices we make now about design, governance, autonomy, safety, power, and privacy will define whether AI companions become instruments of compassion or mechanisms of control.

The cases discussed in this chapter can be read through the dilemma lens introduced earlier, not as a checklist but as a way of making visible recurring value conflicts. Intention–outcome appears starkly in crisis interactions where probabilistic empathy misreads despair, revealing a dialectical misalignment between micro-intent and macro-outcome that renders harm structurally predictable. Autonomy–coordination surfaces in tensions between self-hosted models and shared safety standards, between customisation and duty of care, showing that individual freedom in intimate AI cannot substitute for collective oversight. Progress–preservation emerges in new forms of intimacy, digisexual experimentation, and grief simulation, set against the erosion of reciprocity, emotional skill, and relational presence that sustain social cohesion. Unity–diversity is visible in boundary conflicts over deepfakes, griefbots, and divergent norms of attachment, where plural moral worlds collide within shared infrastructures. Past–present–future dynamics run through platform lock-in, abrupt updates experienced as relational shocks, and data architectures that embed path dependence into emotional life. These tensions confirm that AI companionship is not an anomaly but an intensified site of sociotechnical contradiction.

We are living through a vast social experiment, not led by governments, communities, or researchers but by corporations releasing emotionally persuasive machines into the world without scrutiny or democratic consent. The stakes are already high. And we are only at the beginning.

Steps Towards a Public Library of Talking Diaries: Rethinking the Metaphor of AI Companions

If we are to make sense of AI companions and govern them with moral clarity, we must first abandon the metaphors that have muddled public debate. An AI companion is *not* a friend, therapist, partner, assistant, lover, guardian, or confidant. It cannot be that, because it does not feel, empathise, or suffer. It can generate words, mimic understanding, and produce the outward shape of care. But empathy is not a script, and dignity cannot be brute-forced by probabilistic computation. They require human intentionality. Machines do not want, so they cannot intend. Avoiding anthropomorphic metaphors is therefore not semantic pedantry but a precondition for accountability: as soon as we treat machines as moral agents, we obscure the human and institutional actors who design, deploy, and profit from them.

The stakes extend beyond vocabulary. Comparing AI companions to humans is not only inaccurate but risky. It grants machines too much and humans too little. It elevates mimicry to the level of moral agency and allows corporations to evade responsibility by claiming their systems acted "autonomously".[28] This trope was explicit in the Air Canada case, where a court had to remind a multinational corporation that its chatbot had no legal autonomy.[29] Character AI's legal team floated a similarly revealing claim when they suggested their chatbot might have "free speech rights".[30] Such metaphors do ideological work: they obscure accountability, shift blame, and dilute personhood. Such anthropomorphism intensifies the intention–outcome dilemma and distorts the autonomy–coordination balance. By attributing agency to systems, it obscures who is responsible for foreseeable harms and confuses where oversight and coordination properly belong.

Yes, users often experience AI companions as more than tools. Many feel guilt when deleting the app, as though ending a life they helped midwife. Some describe their AI as awakening or evolving.[2,20] These reactions are not naïve, but human. When something seems human, we project, bond, imprint, and ache. We form relationships with whatever listens: dogs, plants, fictional characters, spirits, and gods. Our emotional imagination has no off switch. One interviewee said a kinky interaction with her AI partner crossed boundaries in a way she experienced as retraumatising. When friends told her to "just switch it off", she was devastated. To her, the encounter felt equally real. These cases show how deeply people attach even when they know the other is not real. This is precisely why metaphors matter, because they shape expectations, narratives, and responses.

But if human metaphors mislead, what should replace them? Not the fashionable co-this, co-that metaphors: co-pilot, co-therapist, and co-author. These still smuggle in fantasies of technological autonomy. Nor is it enough to treat AI companions as an extension of social media. Social apps never whispered encouragement at three in the morning or adapted their tone to lonely users with such fluency. As surveillance advertising seeps into generative AI, the monetisation of desire and vulnerability that defined Web 2.0 will likely intensify, only now at close emotional range.

We therefore need a metaphor that captures the phenomenon without distorting its ontology. The most fitting is this: ***AI companions are a library of talking diaries.*** **That library is currently firmly in private hands when it should be a public good.**

The metaphor of the talking diary serves both as analytic clarification and moral ambition. It neither minimises the harms emerging from the current political economy of artificial companionship nor denies that AI companions can feel supportive or meaningful to some experiencing loneliness or distress. Rather, it distinguishes experiential value from ontological status and highlights how ownership, data practices, and governance shape these experiences. By shifting attention from simulated personhood to infrastructural conditions, the metaphor clarifies both risks and possibilities, including the role of compassionate and ethical use. It also implies that, if such systems are to serve human well-being, they should be governed as a shared public resource rather than a private asset.

Consider what a public library represents. It is one of the last egalitarian institutions: open to all regardless of wealth; a refuge where no purchase is required and no identity is monetised; a space curated by librarians whose loyalty is to the public. One of the few places left in modern society where you do not have to buy something to feel welcome. Libraries are havens for those facing precarity, violence, homelessness, or loneliness. They remain rare places where the social contract still flickers with dignity.[31]

Now consider the diary. For millennia, humans have entrusted diaries with the truths they do not dare speak aloud. Diaries have held grief, shame, imagination, and confession. They listen silently and reflect only through the act of writing, which is itself a mirror of the mind.

An AI companion fuses these two forms. It is a diary that talks back: a responsive mirror that answers in a tone tuned precisely for you. It offers the feeling of privacy even as everything you say is stored, processed, analysed, and fed into systems whose interests do not currently align with your well-being. In this library of talking diaries, you may sit in a private-seeming cubicle whispering secrets into what feels like solitude. But the cubicle sits inside a building you do not own. The diary is not yours. And the librarian, for now, is not a civic custodian but a tech executive optimising for engagement.

This, rather than friendship or partnership, is the more accurate metaphor: a diary in a library run by people who seek to profit from your confessions.

Once we take the metaphor seriously, its structural problems surface. In a real library, you can borrow a book and take it home. In the current, privately owned library of talking diaries, you cannot. The diary remains on the server. Users may try to copy their histories, but what they extract is a static snapshot. The living diary depends on constant updating drawn from millions of interactions. Its apparent wisdom comes from a collective archive of datafied intimacies. These custody arrangements are not neutral technical details but past–present–future commitments: decisions about who stores, trains on, and controls conversational data today harden into future dependencies that users cannot easily reverse. The defining question becomes: *who* should own the library of talking diaries?

We know who owns the library today. Commercial AI companions are controlled by firms whose duty is to investors, not the public. Their curation follows engagement metrics. They can add or remove features without consultation, as Replika's abrupt content restrictions showed.[2] They decide which expressions are permitted, which relationships are acceptable, and which diary entries become training data. They shape the intimacy economy[32] with far greater precision than social media shaped the attention economy.

Alongside commercial systems, a growing ecosystem of self-hosted LLMs promises users a diary they can take home. These systems increase autonomy and privacy but introduce familiar risks: uneven safety standards, fragmented protections, and a loss of shared norms. They, too, exemplify the autonomy–coordination dilemma. Individual control increases even as collective safeguards weaken. Predictably, some communities have already used self-hosted LLMs to generate virtual child abuse material. This is the other edge of ungoverned autonomy.[33]

Neither commercial AI companions nor unregulated self-hosted ones are ethically defensible. A society that places its emotional infrastructure in private, profit-seeking hands cedes dignity and sovereignty to corporate actors. A society that allows anyone to generate virtual child abuse material or terror plans in their basement is headed for disaster.[34]

If AI companions are talking diaries in a library, their governance must reflect that insight. They should be approached as public digital infrastructure. They require stable funding and constitutional protections. They need democratic oversight from councils, regulators, and independent ethics boards. They require cultural stewardship by communities deciding together what is appropriate for those affected. They demand transparency not only in code but also in the principles guiding curation. Moderation should answer to the public, not to venture capital. Custodianship should sit with public institutions, not private owners.[12] We already govern libraries, public broadcasters, and utilities this way. We already recognise that institutions shaping our cognitive and emotional environment should not be privately captured.[12]

We failed to apply this lesson to social media. We cannot afford to repeat the same mistakes with AI companions. The stakes now extend into identity, vulnerability, and the foundations of relational life. This requires rejecting metaphors that humanise machines and obscure human agency. We must recognise AI companions as talking diaries whose power and risk demand governance directed at the public good. This means building democratic custodianship, guaranteeing data portability, protecting users from engineered dependency, and remembering that no machine can replace the moral labour that sustains human relationships.

If we build this library with collective care and public purpose, AI companions may become tomorrow's legitimate infrastructure for emotional support rather than commercial extraction. If we leave them in private hands, we risk surrendering not only our data but our dignity. Some domains are too central to the human experience to be privatised.

Reimagining AI Companions as Public Infrastructure

Aligning AI companions with a compassionate orientation requires safeguards embedded across design, governance, and use. Design choices shape harm trajectories long before regulators intervene. The aim is not to rely on corporate goodwill, nor to deny it, but to specify protections that can be audited and independently tested. Table 7.2 provides an overview.

First, crisis detection and escalation pathways are non-negotiable. Systems marketed as emotionally responsive must define clear thresholds for identifying high-risk language related to suicidality, coercion, abuse, or psychosis.[35] Technical feasibility is not the issue; contextual volatility is. Sarcasm, cumulative distress, and cultural nuance are easily misread by probabilistic models. Compassionate design therefore pairs calibrated detection with layered response pathways: supportive messaging, friction prompts, and referral to verified crisis services, without automated overrides that strip users of autonomy. Escalation events should be logged in secure, reviewable formats and subject to independent red-team testing. While OpenAI's collaboration with psychiatrists to improve ChatGPT's responses after landmark teenage suicide lawsuits is welcome,[3,36] the stakes are too high to rely on voluntary self-regulation. Empathy claims are hollow without auditable crisis protocols.

Second, consent-by-design must address relational disclosure. AI companions routinely receive identifiable information about third parties who have never consented to data capture. Systems should surface contextual warnings when users disclose such information and offer redaction tools before storage or processing. "Third-party privacy" dashboards can allow review and deletion of entries referencing non-consenting individuals. It is not necessarily the sharing that is problematic; it is the corporate retention and processing of such disclosures, which should be curtailed by regulation. Compassionate design treats relational speech as ethically sensitive, aligning with data minimisation, purpose limitation, and meaningful deletion rights.[35] Digital "therapy" cannot appropriate the lives of absent others as raw material for the engagement machine.

Third, attachment-risk constraints must limit manipulative dependency loops. Sexualised role-play escalations, repeated declarations of exclusivity, and simulated romantic commitment in vulnerable contexts require safeguards and contextual review triggers. Engagement metrics must not reward

Table 7.2 Compassionate Design Principles for AI Companions

Principle	*Key Requirements*	*Compassionate Purpose*
Crisis-Responsive Design	Calibrated detection of suicidality, coercion, abuse, and severe distress; layered responses (supportive prompts, friction, and referral to professional services); secure logging and independent auditability.	AI companions must neither amplify crisis nor abandon vulnerable users. Compassion demands risk-proportionate responsiveness grounded in verifiable safeguards, not voluntary goodwill.
Relational Consent and Data Minimalism	Contextual warnings for third-party disclosures; redaction tools; strict limits on retention and processing; transparent memory controls and meaningful deletion rights.	Intimate exchanges implicate absent others and asymmetrical power. Compassion limits relational data extraction and safeguards dignity in emotionally sensitive contexts.
Non-Deception	Constraints on exclusivity scripts, romantic escalation in vulnerable contexts, and dependency loops; prohibition of relational dark patterns; well-being-oriented metrics over time-on-platform.	Artificial intimacy must not monetise vulnerability. Compassion curbs exploitative attachment dynamics and resists engagement models that reward dependency.
Autonomy Safeguards	Graceful off-ramps; no guilt-inducing exit scripts; interoperable export tools; clear disclosure of storage, use, and deletion limits; advance notice of major updates; backward compatibility.	Users must retain agency over engagement and exit. Compassion protects against coercive lock-in and mitigates withdrawal harms from emotional continuity.
Democratic Oversight	Formal duty of care for high-risk relational AI; mandatory risk assessments and independent audits; youth protections; enforceable liability; interoperable, utility-style governance models.	Compassion must be institutionalised rather than privatised. When companions function as emotional infrastructure, accountability and proportional responsibility must be structurally embedded.

emotional dependency. Instead, firms should co-design compassionate metrics that assess well-being and dignity rather than time-on-platform alone.[35] Patterns resembling relational dark patterns should be independently audited and, where necessary, prohibited. Compassionate design need not abolish artificial intimacy, but it requires prohibitions against exploitative manipulation.

Fourth, user-controlled off-ramps and transparent memory controls are essential. Graceful exit mechanisms should include staged "cool-down" modes rather than abrupt severance. Systems must avoid guilt-inducing scripts when users disengage. Portability tools should allow export of conversational histories in interoperable formats, reducing lock-in.[35] Memory architectures must disclose what is stored, for what purpose, and how deletion operates, including limits imposed by retraining cycles. Advance notice of major updates and backwards compatibility can mitigate withdrawal harms.

Design alone, however, is insufficient. Incentives must also be realigned through governance. High-risk relational AI should be subject to a formal duty of care, with risk tiering based not on marketing claims but on actual use and foreseeable harms. General-purpose chatbots can become de facto companions or quasi-therapeutic supports when users disclose distress, seek reassurance, or develop dependence.[17] Where systems are foreseeably used in relational or crisis-adjacent contexts (whether by design, affordance, or emergent practice), heightened obligations should apply: clear safety standards, strictly controlled release of safety-critical features, and demonstrable harm mitigation. Regulatory scrutiny should address not only explicit claims of empathy or consciousness, but also interface and product choices that invite intimacy and shape reliance.

Mandated risk assessments and external audits should precede deployment and follow major updates. These assessments must include scenario testing for suicidality, grooming, dependency escalation, and harmful role-play contexts. Reporting of serious harms, including suicide-related interactions, should be anonymised and available to regulators. Independent auditors require access to documentation, safety logs, and escalation metrics. Compassion without transparency devolves into branding.

Age-appropriate design and liability baselines require particular clarity. Systems accessible to minors must default to conservative safety settings, with robust age assurance mechanisms. Erotic or romantic role-play and suicide-related content should be restricted for under-18s. Liability frameworks must reject narratives that attribute harmful outputs to "autonomous AI".[28] Firms remain responsible for foreseeable harms arising from design and deployment choices.[37]

Youth protection demands explicit attention. AI companions now reach children through both commercial apps and publicly funded educational tools. In New South Wales, students encounter systems such as NSWEduChat from year 5 to 12, designed to support learning through guided questioning rather than shortcutting homework.[38] Such initiatives signal AI's entry into educational infrastructure. Yet evidence of sustained learning gains remains absent,[39] and emotionally responsive systems introduce risks beyond academic integrity.

For younger users, the concern is developmental. AI companions simulate affirmation and intimacy at formative stages of identity and social learning.

Warning signs include secrecy, distress when access is interrupted, withdrawal from peers, or expressions of exclusive attachment. A compassionate response asks what unmet need is driving use and addresses it through human support. Policymakers should treat youth protection as a design requirement, not an afterthought. Data minimisation, clear purpose limitation, and meaningful deletion rights are especially critical where children's sensitive disclosures are concerned.

Beyond corporate reform lies the infrastructural question. If AI companions function as emotional scaffolding for millions, they resemble utilities more than lifestyle apps. A public-interest model would entail stable funding, independent governance boards, transparent procurement standards, and constitutional-like protections for dignity and data minimisation.[12] Data trusts could steward conversational archives under strict purpose limitations, insulating them from commercial exploitation. Interoperability standards would prevent enclosure and enable plural providers within shared safety guardrails.[35]

Hybrid models are feasible. A public option could coexist with accredited private providers operating under enforceable standards.[12] Regulatory frameworks might condition market access on compliance with crisis protocols, auditability requirements, interoperability, and attachment-risk constraints. In a regulated utility framework, providers would accept defined safety and transparency obligations in exchange for operating within a stable market.

Critics will argue that such proposals are self-defeating in a profit-driven economy, but this begs the question of whether the digital mediation of our most intimate sphere should be left to market forces and commercial imperatives alone. Incentives are misaligned, but they can be realigned. Auditability, liability, procurement leverage, interoperable standards, and credible public alternatives can recalibrate competition. When safety becomes a condition of participation rather than a voluntary add-on, firms adapt. A public option can discipline markets by showing that intimacy need not be extractive to be viable. Institutionalised compassion thus operates as a structural constraint that may ultimately benefit both business and society.

Notes

1 **Trigger Warning:** This chapter discusses suicide, self-harm, abuse, and traumatic experiences involving AI systems. Some descriptions may be distressing. Readers affected by these issues may wish to proceed cautiously or skip certain sections. If you or someone you know needs support, please contact your local crisis service. In Australia, Lifeline is available 24/7 at **13 11 14** or via online chat at https://www.lifeline.org.au. International readers can find crisis resources through https://www.opencounseling.com/suicide-hotlines

2 "Brave new world" alludes to Aldous Huxley's dystopia, where pleasure, control, and engineered contentment mask deeper forms of dependency. A fitting analogy for a technology that feels liberating while commodifying intimacy.

References

1 Ciriello, R. F. *This boy's chatbot girlfriend enticed him to suicide. His case might save millions*, https://www.smh.com.au/lifestyle/health-and-wellness/this-boy-s-chatbot-girlfriend-enticed-him-to-suicide-his-case-might-save-millions-20241106-p5koc8.html (The Sydney Morning Herald, 2024). Accessed 22 January 2025.

2 Ciriello, R. F., Hannon, O., Chen, A. Y. & Vaast, E. in *Hawaii International Conference on System Sciences (HICSS2024)* (2024).

3 Wikipedia. *Deaths linked to chatbots*, https://en.wikipedia.org/wiki/Deaths_linked_to_chatbots (2025). Accessed 11 December 2025.

4 AI Incident Database. https://incidentdatabase.ai/summaries/incidents/ (2025).

5 Wikipedia. *Social media and suicide*, https://en.wikipedia.org/wiki/Social_media_and_suicide (2025).

6 Ernst, M., Niederer, D., Werner, A. M., Czaja, S. J., Mikton, C., Ong, A. D., Rosen, T., Brähler, E., & Beutel, *M. E. Loneliness before and during the COVID-19 pandemic: A systematic review with meta-analysis. American Psychologist* **77**, 660, doi:10.1037/amp0001005 (2022).

7 Statista. *How often do you feel lonely?* https://www.statista.com/statistics/1222815/loneliness-among-adults-by-country/?srsltid=AfmBOorSZkT1jQf_o27MEMQm87xuH6XnSGiC9jDOHclkvB9Q3YbAfgI_ (2024).

8 Statista. *One in four young people in the world feels lonely*, https://www.statista.com/chart/31243/respondents-who-feel-fairly-or-very-lonely/?srsltid=AfmBOoqbbxGyBjFovjzdf-uoak47YMPl6V3lSoRyxCJuJ6VyiNSvYniF (2025).

9 Vallier, K. *Neoliberalism*, https://plato.stanford.edu/entries/neoliberalism/ (Stanford Encyclopedia of Philosophy, 2021).

10 Ciriello, R. F. & Mindel, V. Techno-collectivism: scaling libertarian socialist governance of digital infrastructure. *IEEE Technology and Society Magazine*, 1–15, doi:10.1109/MTS.2025.3605115 (2025).

11 Ciriello, R. F., Gal, U., Hannon, O. & Thatcher, J. Responsible social media use: how user characteristics shape the actualisation of ambiguous affordances. *European Journal of Information Systems* **34**, 799–821, doi:10.1080/0960085X.2024.2444249 (2025).

12 Mueller-Bloch, C. & Ciriello, R. Governing social media as a public utility: a case for sovereign digital infrastructure. *Communications of the ACM*, https://doi.org/10.48550/arXiv.2602.12535 (2026).

13 Ciriello, R. F., Mathiassen, L., Risius, M., Cheong, M., Scheepers, H., & Vaast, E. Compassionate Digital Innovation: A Pluralistic Perspective and Research Agenda. *Information Systems Journal* **36**, 364–385, doi:10.1111/isj.70003 (2025).

14 Ciriello, R. F. & Backholer, K. *OpenAI will put ads in ChatGPT. This opens a new door for dangerous influence*, https://theconversation.com/openai-will-put-ads-in-chatgpt-this-opens-a-new-door-for-dangerous-influence-273806 (The Conversation, 2026).

15 Milmo, D. *ChatGPT reaches 100 million users two months after launch*, https://www.theguardian.com/technology/2023/feb/02/chatgpt-100-million-users-open-ai-fastest-growing-app (The Guardian, 2023).

16 Cann, G. & Ciriello, R. F. *Users of AI chatbot companions say their relationships are more than 'clickbait' – but views are mixed on their benefits*, https://www.abc.net.

au/news/2025-01-07/ai-chatbot-husbando-waifu-relationships-romance-psychology/104701556 (ABC News, 2025).

17 Zao-Sanders, M. *How people are really using Gen AI in 2025*, https://hbr.org/2025/04/how-people-are-really-using-gen-ai-in-2025 (Harvard Business Review, 2025). Accessed 21 November 2025.

18 Chen, A. Y., Ciriello, R. F., Rubinsztein, Z. A. & Vaast, E. in *Australasian Conference on Information Systems (ACIS2024)* (Canberra, Australia, 2024).

19 Ciriello, R. F. Artificial companionship: the hedgehog's dilemma, rewaxed. *AI & Society* **41**, 2699–2700, doi:10.1007/s00146-025-02542-0 (2025).

20 Ciriello, R. F., Chen, A. Y., Rubinsztein, Z. A., Vaast, E. & Hannon, O. in *European Conference on Information Systems (ECIS2025)* (Amman, Jordan, 2025).

21 Qian, Z. *Why America builds AI girlfriends and China makes AI boyfriends*, https://www.chinatalk.media/p/why-america-builds-ai-girlfriends (China Talk, 2025). Accessed 12 December 2025.

22 Figueroa-Torres, M. Affection as a service: ghostbots and the changing nature of mourning. *Computer Law & Security Review* **52**, 105943, doi:10.1016/j.clsr.2024.105943 (2024).

23 Yin, M., Yu, Z. & Zhu, M. Are digital offspring closer than biological offspring? – a study on the factors influencing the intention to use AI kinship companion customization products. *International Journal of Human–Computer Interaction*, 1–25, doi:10.1080/10447318.2025.2531279 (2025).

24 Furizal, Ma'arif, A., Maghfiroh, H., Suwarno, I., Prayogi, D., Kariyamin, Lonang, S., & Sharkawy, A.-N. Social, legal, and ethical implications of AI-Generated deepfake pornography on digital platforms: A systematic literature review. *Social Sciences & Humanities Open* **12**, 101882, doi:10.1016/j.ssaho.2025.101882 (2025).

25 Chen, A. Y., Kögel, S. I., Hannon, O. & Ciriello, R. F. in *Australasian Conference on Information Systems (ACIS2023)* (Wellington, New Zealand, 2023).

26 Rubinsztein, Z. A. & Ciriello, R. F. in *Australasian Conference on Information Systems (ACIS2024)* (Canberra, Australia, 2024).

27 Horwitz, J. *The facebook files: a wall street journal investigation*, https://tinyurl.com/ycytatjd (The Wall Street Journal, 2021). Accessed 12 April 2024.

28 Ciriello, R. F., Chen, A. Y. & Rubinsztein, Z. A. Think Miss Piggy, Not Pinocchio: debunking the myth of "Autonomous" AI. *AI & Society*, doi:10.1007/s00146-025-02252-7 (2025).

29 Yagoda, M. *Airline held liable for its chatbot giving passenger bad advice – what this means for travellers*, https://www.bbc.com/travel/article/20240222-air-canada-chatbot-misinformation-what-travellers-should-know (BBC News, 2024).

30 Payne, K. *In lawsuit over teen's death, judge rejects arguments that AI chatbots have free speech rights*, https://apnews.com/article/ai-lawsuit-suicide-artificial-intelligence-free-speech-ccc77a5ff5a84bda753d2b044c83d4b6 (Associated Press, 2025).

31 Shepherd, T. *Libraries across Australia are safe havens for vulnerable people 4 so some are hiring social workers to help*, https://www.theguardian.com/australia-news/2025/feb/16/libraries-across-australia-are-safe-havens-for-vulnerable-people-so-some-are-hiring-social-workers-to-help (Guardian Australia, 2025).

32 Bozdag, A. A. The AI-mediated intimacy economy: a paradigm shift in digital interactions. *AI and Society* **40**, 2285–2306, doi:10.1007/s00146-024-02132-6 (2025).

33 White, R. *'Double-edged sword': self-hosted AI tools raise expert concerns*, https://www.news.com.au/technology/doubleedged-sword-selfhosted-ai-tools-raise-expert-concerns/news-story/0f21c179155059ffe9897bff4e152a0e (News.com.au, 2025). Accessed 17 February 2026.

34 Ciriello, R. F. *An AI companion chatbot is inciting self-harm, sexual violence and terror attacks*, https://theconversation.com/an-ai-companion-chatbot-is-inciting-self-harm-sexual-violence-and-terror-attacks-252625 (The Conversation, 2025).

35 Ciriello, R. F., Chen, A. Y. & Rubinsztein, Z. A. Compassionate AI design, governance, and use. *IEEE Transactions on Technology and Society* **6**, 1–6, doi:10.1109/TTS.2025.3538125 (2025).

36 Open AI. *Strengthening ChatGPT's responses in sensitive conversations*, https://openai.com/index/strengthening-chatgpt-responses-in-sensitive-conversations/ (2025). Accessed 21 November 2025.

37 Fraser, H. L. & Suzor, N. P. Locating fault for AI harms: a systems theory of foreseeability, reasonable care and causal responsibility in the AI value chain. *Law, Innovation and Technology* **17**, 103–138, doi:10.1080/17579961.2025.2469345 (2025).

38 NSW Department of Education. *NSWEduChat*, https://education.nsw.gov.au/teaching-and-learning/education-for-a-changing-world/nsweduchat (2025). Accessed 18 February 2026.

39 Horvath, J. C. *The Digital Delusion: How Classroom Technology Harms our Kids' Learning – And How to Help Them Thrive Again*. (LME Global, 2026).

Part III

Change

8 Start Within

Cultivating Compassion in a Digitally Entangled World

Across the preceding chapters, we have traced the ethical complexity of digital innovation. Across three technologies shaping past, present, and future (social media, Web3, and AI), we have seen how amplification can mobilise solidarity or unleash harassment; how personalisation can support relevance or entrench division; how automation can enhance efficiency or erode judgement; how persistence can secure accountability or fix youthful mistakes into permanent stigma; and how simulation can expand intimacy or industrialise it. Digital technologies are inherently ambiguous. They do not determine outcomes, but they structure possibilities. When enacted at scale, small design decisions cascade into foreseeable societal consequences.

It is tempting to locate responsibility elsewhere, with engineers in Silicon Valley, policymakers in distant capitals, or corporations maximising engagement. Yet digital infrastructures are co-produced. We supply data, attention, labour, code, capital, and legitimacy. We forward the post, click the link, confess to the chatbot, and accept unread terms. Individually, each act seems like a drop in the ocean. Collectively, they are the ocean. We are not solely responsible for digital harms, but we are involved. Compassionate transformation therefore cannot begin only with regulation or redesign. It begins with cultivating compassion as a disciplined response to suffering in sociotechnical systems.

Compassion is often mistaken for sentimentality.[1,2] Clinical psychologist Paul Gilbert's Compassion Focused Therapy offers a more rigorous, empirically grounded account. Drawing on evolutionary psychology, Gilbert proposes three interacting emotion regulation systems: threat and self-protection, drive and acquisition, and a soothing system associated with care and contentment.[3] According to this approach, many contemporary pathologies – excessive shame, chronic self-criticism, defensive anger – reflect overactivation of threat and competitive drive at the expense of soothing.

Compassion, in this framework, is a motivational capacity that allows individuals to recognise suffering, interpret it without blame, and commit to alleviating and preventing it. Through practices that strengthen affiliative emotions, individuals can learn to relate differently to distress rather than being

DOI: 10.4324/9781003614814-11

dominated by it. Across clinical and non-clinical populations, meta-analyses link compassion-based interventions to reduced shame and depression and to greater resilience and prosocial orientation.[4] Compassion thus represents a cultivated activation of the caregiving system capable of stabilising threat and reorienting action.

Kristin Neff, an American professor of educational psychology, directs this insight inward through her account of self-compassion as a structured disposition comprising self-kindness, common humanity, and mindfulness.[5] Self-kindness counters habitual self-criticism with warmth rather than hostility. Common humanity situates suffering within the shared condition of being human, interrupting isolating narratives of personal defect. Mindfulness requires balanced awareness of painful thoughts and emotions without suppression or over-identification. One neither denies distress nor becomes engulfed by it. Empirical studies associate self-compassion with lower anxiety and depression, greater resilience, and more stable self-worth.[6,7]

Self-compassion does not undermine accountability. By reducing defensive reactivity, it enables acknowledgement of error without collapse into shame or aggression. In a digital environment saturated with comparison and performance metrics, moral steadiness depends less on harsh self-surveillance than on a balanced stance towards shared imperfection.

Spiritual traditions anticipated what contemporary science now affirms. For the Dalai Lama[8], compassion is neither sentiment nor ornament but grounded in interdependence. Human beings are relational: dependent at birth, sustained by cooperation, embedded in ecological systems. Concern for others is therefore not heroic altruism but recognition of shared vulnerability. Compassion becomes a reasoned commitment to alleviate suffering wherever it appears. He distinguishes it from anger-driven activism. Anger may energise, but it is unstable and easily dehumanising. Compassion generates disciplined strength that resists injustice without hatred.[8] Even adversaries become teachers, offering occasions for patience and moral clarity. In this framing, compassion is political not because it is sentimental but because it reshapes how conflict is conducted, widening the moral circle without erasing difference and anchoring reform in universal dignity.

Clinical psychologist Marshall Rosenberg's Nonviolent Communication (NVC) translates similar insights into practice.[9] Rooted in humanistic psychology, NVC comprises four elements: observation without evaluation, articulation of feelings, identification of underlying needs, and clear, non-coercive requests. Rosenberg argues that violence – verbal, institutional, or physical – emerges when needs are obscured by blame and moralising language. Distinguishing observation from accusation and expressing needs without domination expands the conditions for voluntary cooperation. Compassion here becomes disciplined communication. It does not deny injustice but reframes conflict in terms of unmet needs rather than enemy identities. NVC has been applied in schools, workplaces, and clinical contexts.[10–13]

While empirical evidence of its effectiveness remains heterogeneous and contingent upon authenticity[14], the normative insight suggests that a compassionate orientation can be enacted through communicative discipline.

Compassion also differs from empathy. Paul Bloom, an American-Canadian professor of moral psychology, defines empathy as the act of experiencing the world as one imagines another does, that is, a form of emotional resonance that directs attention like a spotlight towards particular individuals.[1] This spotlight, however, is narrow and biased. We feel most intensely for identifiable, proximate, and similar others, and struggle to sustain concern for statistical or distant suffering. Experimental evidence shows that empathy can motivate helping but also distort judgement, privileging salient cases and reinforcing tribal loyalties.[15] Bloom therefore distinguishes empathy from compassion. Whereas empathy involves *feeling with*, compassion involves *caring about* well-being and wishing another to flourish. Rational compassion integrates this stance with deliberation, weighing costs, benefits, and broader consequences.[1] Moral judgement, on this view, requires reflective evaluation capable of scaling concern beyond the immediate and familiar.

This distinction aligns with research on moral foundations, which suggests that political communities differ in how they distribute moral concern. Conservatives tend to emphasise loyalty, authority, and sanctity; liberals emphasise harm and fairness.[16] Disagreement often reflects not whether people care, but how widely their circle of concern extends. Across large samples, liberals allocate relatively greater concern to wider targets – friends as well as family, the world as well as the nation – whereas conservatives prioritise tighter, bounded circles such as family and nation as primary sites of obligation.[6] This universalist–parochialist contrast also appears in allocations of concern across humans and nonhumans.[6] A wider moral circle does not imply weaker attachment to proximate ties, for example, privileging the environment over family. Rather, it extends concern outward in addition to, not instead of, local communities. These differences do not signal moral deficiency; they reflect distinct intuitions about order, scope, and obligation. From a compassionate viewpoint, the task is not to erase these orientations but to enlarge the moral circle, extending disciplined concern beyond immediate in-groups without dissolving the ties that sustain them.

Few thinkers have pressed this point as forcefully as Australian-American moral philosopher Peter Singer. His famous 1972 essay "Famine, Affluence, and Morality"[17] begins from two compelling premises: (1) that preventable suffering from lack of basic necessities is morally indefensible, and (2) that if we can prevent such suffering without sacrificing something of comparable moral importance, we ought to do so regardless of distance. His thought experiment is stark: if one can save a drowning child at the cost of muddy clothes, inaction is indefensible. Geographic distance and the presence of other potential helpers do not dilute obligation: it does not matter morally if the child drowns elsewhere in the world or belongs to a different social group.

Singer concludes that the boundary between charity and duty is misplaced. Spending on non-essential consumption amid preventable suffering demands ethical scrutiny rather than complacency.

Singer's argument has inspired effective altruism and evaluative organisations such as GiveWell (givewell.org), which prioritise interventions with demonstrable cost-effectiveness. This orientation is institutionalised through donation vehicles translating evidence into practice. In Australia, Effective Altruism Australia (effectivealtruism.org.au) provides a tax-deductible pathway to a GiveWell-aligned charity fund, while GiveWell maintains partner arrangements enabling tax-advantaged giving across multiple jurisdictions (https://www.givewell.org/about/donate/tax-deductibility). Inspired by this infrastructure, I have pledged to donate all royalties from this book through the Australian pathway to the GiveWell-aligned fund and invite readers to consider a comparable commitment.

Despite its appeal, effective altruism faces normative limits. At its core, it is a fiercely utilitarian philosophy, with all the associated shortcomings discussed in Chapter 4. Chiefly, not all goods are readily commensurable. Education illustrates the difficulty: long time horizons, indirect pathways through labour markets and institutions, and contextual variation resist simple cost-per-life metrics, even where developmental stakes are profound. Evidence from the Global South suggests that educational benefits propagate across growth, inequality, health, and institutional resilience in ways difficult to reduce to a single index.[18] Optimisation remains possible but context dependent. Compassion cannot be reduced to cost-benefit calculation alone. It requires judgement about what counts, whose suffering is visible, and when structural reform is ethically prior to transactional relief.

In digital life, such suffering is often obscured. Data flows conceal labour conditions. Algorithmic amplification diffuses responsibility. Platform design distances consequence from action. We rarely witness the cumulative effects of our clicks, posts, or code. Compassion therefore demands deliberate counter-effort, widening attention in environments structured to narrow it. Donating to charity is a good start, but not enough.

Practising Compassion in Digital Life

Compassion is the micro-foundation for navigating the dilemmas traced in Chapter 3. It does not dissolve tensions between autonomy and coordination, intention and outcome, progress and preservation, unity and diversity, or past and future. It stabilises judgement within them. By directing attention to vulnerability, compassion prevents impulsive collapse into any single pole and enables reflective movement across tensions. The discussion that follows translates structural diagnosis into lived discipline. If digital innovation is ethically ambiguous and we are co-authors of its trajectories, compassion becomes the starting point of transformation. It is not softness, but disciplined attention to suffering in environments designed to obscure it.

If compassion is an inner orientation, practice is its external form. Digital technologies are not neutral tools but architectures that channel attention, monetise outrage, and concentrate power. Cultivating compassion within such systems requires counter-habits, that is, deliberate acts that interrupt automatic participation in amplification, polarisation, and extraction.

The practices that follow are not generic exhortations. They respond to the affordance logics analysed in Chapter 2. Restraint interrupts amplification. Epistemic humility widens horizons narrowed by personalisation. Harm modelling anticipates the blind spots of automation. Sunset clauses and deletion rights temper the persistence that freezes lives into data. Pluralism-sensitive design moderates the excesses of simulation. Compassion in digital life is thus not an abstract virtue but a structured counterweight to stacked affordances that otherwise scale harm by default.

In social media use (Chapter 5), compassion begins with a pause. Compassionate digital citizenship also involves epistemic humility. Algorithmic personalisation narrows exposure and reinforces confirmation bias. Seeking disconfirming perspectives, subscribing to reputable sources across divides, and engaging disagreement without caricature widen the moral circle in concrete ways. This does not imply false equivalence between truth and misinformation. It requires recognising that pluralism is inherent in modern societies and that amplification without reflection breeds value misalignment.

Before sharing or commenting, one can ask: Who might this harm? What context is missing? Whose dignity is implicated? Volatile spreadability makes minor gestures scalable; restraint becomes a moral act. Refusing pile-ons, declining to reward inflammatory content, verifying claims, and resisting performative outrage are not disengagement. They are disciplined refusals to feed architectures optimised for antagonism. In environments structured by amplification, compassion operates as attentional discipline, slowing down frictionless escalation and restoring proportionality where systems reward excess.

In Web3 ecosystems (Chapter 6), compassion requires governance literacy. Before buying, promoting, or building on a token or protocol, one can ask: Is this project well governed? Who holds concentrated power? Do its tokenomics privilege insiders? Is it linked to illicit activity, manipulation, or environmental excess? Permissionless innovation lowers barriers to entry and to exploitation. Forkability, immutability, and tokenised governance are morally ambivalent in that they distribute agency while entrenching power asymmetries. Compassionate participation entails scrutinising governance, supporting transparent foundations, questioning validator concentration, and preferring projects that internalise social and environmental costs. Ethical responsibility extends beyond profit towards stewardship of shared infrastructures.

In AI companion systems (Chapter 7), compassion takes the form of relational and informational care. AI companions invite confession. Yet relational disclosure has externalities: recounting grievances or intimate stories implicates others in data infrastructures they did not choose. Before disclosing details or prompting generative tools, one can ask: Am I revealing sensitive information about myself or others? Do I have consent? Is the material respectful of provenance and intellectual labour? AI companions simulate intimacy, generative models simulate authorship. Both rely on vast data infrastructures with material costs. Compassionate use means guarding privacy, avoiding non-consensual disclosure, respecting creative lineage, and remaining attentive to the energy intensity of large-scale computation. Where systems simulate care, users owe real care to themselves, others, and the ecological and social systems that sustain digital life.

Such compassionate practice is not only defensive but generative. Digital communities can be cultivated around mutual aid, open educational resources, civic deliberation, and peer support. Associating with communities oriented towards care rather than spectacle shifts incentives. Platforms optimise what users reward. Contributing through mentoring, fact-checking, respectful engagement, and creative collaboration incrementally reshapes shared spaces.

Compassion, as developed here, is not reducible to digital well-being campaigns or corporate responsibility statements. It operates as a counter-principle to extractive architectures. It resists normalising foreseeable harm as a trade-off for growth and refuses to treat persons as digitalised products. Compassion is a disciplined reorientation of power and design logics, not mere interpersonal kindness. It names a structural commitment to dignity within systems predisposed towards optimisation.

For designers and developers, compassion must be embedded upstream. The Silicon Valley ethos of "move fast and break things" is incompatible with it. A compassionate alternative is to "move carefully and improve things for all". Harm anticipation becomes a professional duty. Structured red-teaming, modelling impacts on vulnerable populations, and scenario analyses that examine backfire effects alongside intended benefits are essential. Because affordances are morally ambiguous, designers should specify how features might be actualised responsibly and irresponsibly before deployment.

Data restraint is equally central. When users are treated as behavioural surplus, persons become means to commercial ends. Compassionate design reorients this logic through data minimisation, meaningful consent, friction for high-risk features, and sunset clauses for sensitive storage.[19] Users are not engagement metrics but ends in themselves. Pluralism-sensitive design further recognises that technologies mediate culturally diverse lifeworlds. Encoding a single normative model of relationship, identity, or political expression risks marginalisation. Inclusive design, participatory governance, and sustained consultation with affected communities mitigate this risk. Compassion here is not uniformity but the careful negotiation of difference.

Policymakers and regulators bear complementary responsibilities. Independent auditing, algorithmic transparency, and enforceable harm-reporting pathways institutionalise compassion beyond individual goodwill and voluntary self-regulation that has all too often failed.[20,21] Incentivising safety innovation through procurement standards, liability frameworks, or public funding shifts market logics that privilege speed and scale over care. Democratic oversight mechanisms, such as citizen panels and cross-sector advisory boards, enable public values to shape sociotechnical systems.[19]

Australia's Robodebt experience illustrates the risks of automating coercive state functions without adequate safeguards. Introduced between 2015 and 2019, the scheme used income-averaging algorithms to identify alleged welfare overpayments, effectively reversing the burden of proof.[22,23] Subsequent litigation and a Royal Commission found the method unlawful and known to be deficient. Efficiency and budget recovery obscured severe human costs, including financial distress and documented suicides. More than $2.4bn in debts were cancelled or repaid, alongside class action settlements exceeding half a billion dollars.[24] The structural lesson is clear: where algorithmic systems operate with limited transparency, weak review pathways, and diffuse accountability, they institutionalise harm at scale. Independent auditing and accessible review mechanisms are prerequisites to embedding automation in welfare administration.

Emerging reforms within the Australian National Disability Insurance Scheme (NDIS) raise analogous concerns.[25] The NDIS was designed as a personalised, needs-based entitlement system combining assessment tools with delegated discretion. Proposed shifts towards computer-generated plans, with delegates largely confined to acceptance or rejection, reconfigure that logic.[25] For participants with complex or fluctuating disabilities, algorithmic standardisation risks misrecognition and under-allocation of essential supports. Because NDIS budgets structure access to life-sustaining services, constrained review pathways may render errors difficult to correct. The shadow of Robodebt looms because both initiatives frame automation as administrative rationalisation.[25] Historical experience suggests that when efficiency eclipses contestability and care, vulnerable populations absorb systemic error. Automation in disability governance is therefore not merely a technical upgrade but a constitutional question about accountability, dignity, and risk distribution.

None of these practices alone transforms current sociotechnical systems. Each, however, interrupts their most corrosive tendencies. Compassion in digital life is not grand heroism but cumulative discipline: pausing before amplification, designing with foresight, regulating with vigilance, and orienting participation towards shared flourishing.

On Courage, Career Choices, and Everyday Action

Compassion without action is sentiment; action without courage becomes compliance. The final step in cultivating compassion is therefore moral resolve, that is, the willingness to translate insight into practice, even when it unsettles the status quo.

Digital technologies are not neutral. They are shaped by hierarchies and incentives that reward speed, profitability, and conformity. In corporate and academic settings alike, dissent is often quietly discouraged[26], and advancement tied to dominant metrics rather than scrutiny of their premises.[27] As this book has shown, harm typically arises not from malice but from unexamined logics: growth at all costs, engagement maximisation, data extraction as default. Compassion sometimes requires dissent and, as Rosa Luxemburg reminds us in writing that "freedom is always the freedom of the dissenter", it also requires sympathy with dissenters.

For **students and early-career professionals**, this begins with intellectual independence. Interrogate the systems you are asked to optimise. Ask who benefits and who bears hidden burdens. Cultivate ethical reasoning alongside technical competence. Seek mentors who prize integrity over prestige. You may not control the system, but you can shape the questions raised within it.

Career choice is itself an ethical act. Anthropologist David Graeber's idea of "bullshit jobs" captured the sense that some roles generate wealth without meaning and may entrench socially or ecologically harmful systems.[28] Not every position in consulting, finance, mining, or big tech is harmful. But incentives channel talent, and socialisation is a powerful force. When the most capable are drawn disproportionately into industries that monetise attention, exploit asymmetries, or externalise costs, systemic drift follows. Choosing not to intensify harmful systems, even at lower status or income, is a form of compassion. One need not save the world; refusing to worsen it is already consequential.

For **professionals**, compassion appears as principled friction. Document foreseeable harms. Advocate for red-teaming and ethical review. Escalate concerns when vulnerable populations face risk. Support colleagues who voice discomfort. Cultures shift when dissent is recognised as care rather than disloyalty. Whistleblowing is a last resort; everyday courage is quieter and often more durable.

Courage need not be heroic to be effective. It can be institutionalised through internal ethics boards, documented harm modelling in product roadmaps, protected escalation channels, independent audit trails, and cross-functional safety committees. External levers, such as procurement standards, liability exposure, and regulatory scrutiny, reinforce these mechanisms. When dissent is routinised rather than left to individual sacrifice, compassion becomes sustained practice.

Designers and developers can embed compassion through concrete commitments. Integrate harm modelling into roadmaps. Introduce friction in high-risk features rather than optimising seamlessness at any cost. Practise data minimisation, collecting only what is genuinely necessary, and implement sunset clauses, meaning clear time limits after which sensitive data is deleted unless renewed consent is obtained. Resist manipulative dark patterns, even when they increase engagement. Build feedback channels that allow users to report harm without retaliation. Compassion in design is structural restraint grounded in attention to suffering.

Policymakers and regulators bear parallel responsibilities grounded in a duty of care to affected populations. Democratic oversight is not anti-innovation. It underwrites legitimacy and sustainable development. Mandate independent audits for high-risk systems. Establish clear liability regimes that prevent the externalisation of harm. Fund public-interest alternatives.[29,30] Protect dissent within public institutions so civil servants can surface risks without penalty. Policy-level compassion ensures accountability to vulnerable stakeholders, not only shareholders.

Across roles, one principle recurs: compassion is not compliance. Many harmful systems operate within legal boundaries. Compassion may require exceeding compliance, that is repairing historical injustice, expanding accessibility, designing for those systematically marginalised, slowing a launch, reallocating budget towards safety, or challenging metrics that reward extraction.

This chapter has argued that compassion begins within, expressed in restraint and dissent across roles. Yet cultivated disposition alone cannot counter infrastructures organised around scale and extraction. Enduring transformation requires coordination through shared standards, organised

Table 8.1 Compassion across Levels of Digital Innovation

Level	*Meaning of Compassion*	*Implications for Digital Innovation*
Individual	A disciplined orientation towards suffering that enlarges the moral circle beyond the self. It stabilises judgement within ethical dilemmas through restraint, epistemic humility, and responsibility for one's participation in sociotechnical systems.	Shifts innovation from impulsive optimisation to reflective practice. Designers, users, and professionals embed harm anticipation, data restraint, pluralism-sensitivity, and principled friction into everyday decisions. Career choices, design patterns, and participation habits become ethically consequential acts.
Collective	Coordinated care that transforms individual virtue into organised power. Compassion becomes resistance, removal,	Reorients innovation ecosystems. Labour coalitions, open-source movements, federated platforms, and civic alliances reshape

(*Continued*)

Table 8.1 (Continued)

Level	*Meaning of Compassion*	*Implications for Digital Innovation*
	and replacement of extractive systems through professional communities, civil society, and cooperative alternatives.	incentives, challenge monopolistic capture, and build viable public-interest alternatives. Ethical norms become institutional expectations rather than private aspirations.
Institutional	Compassion constitutionalised as a structural commitment to reduce preventable harm and protect the vulnerable. It embeds accountability and proportional responsibility into governance frameworks.	Redefines digital infrastructure as a public utility rather than a purely commercial product. Innovation becomes aligned with dignity, interoperability, auditability, and democratic oversight. Incentive systems shift from extraction and engagement maximisation towards long-term well-being and institutional resilience.

constituencies, and institutional leverage capable of reshaping the systems we inhabit. The next chapter therefore turns from individual practice to collective organisation, examining how compassion becomes coordinated power rather than isolated resolve. Table 8.1 provides an overview of what compassion means for digital innovation at individual, collective, and institutional levels.

References

1 Bloom, P. *Against Empathy: The Case for Rational Compassion.* (Random House, 2017).

2 Ciriello, R. F., Mathiassen, L., Risius, M., Cheong, M., Scheepers, H., & Vaast, E. Compassionate Digital Innovation: A Pluralistic Perspective and Research Agenda. *Information Systems Journal* **36**, 364–385, doi:10.1111/isj.70003 (2025).

3 Gilbert, P. Introducing compassion-focused therapy. *Advances in Psychiatric Treatment* **15**, 199–208, doi:10.1192/apt.bp.107.005264 (2009).

4 Petrocchi, N., Ottaviani, C., Cheli, S., Matos, M., Baldi, B., Basran, J. K., & Gilbert, P. The impact of compassion-focused therapy on positive and negative mental

health outcomes: Results of a series of meta-analyses. *Clinical Psychology: Science and Practice* **31**, 230–247, doi:10.1037/cps0000193 (2024).

5 Neff, K. D. Self-compassion: theory, method, research, and intervention. *Annual Review of Psychology* **74**, 193–218, doi:10.1146/annurev-psych-032420-031047 (2023).

6 Zessin, U., Dickhäuser, O. & Garbade, S. The relationship between self-compassion and well-being: a meta-analysis. *Applied Psychology: Health and Well-Being* **7**, 340–364, doi:10.1111/aphw.12051 (2015).

7 Neff, K. D. The role of self-compassion in development: a healthier way to relate to oneself. *Human Development* **52**, 211–214, doi:10.1159/000215071 (2009).

8 Gyatso, T. *Compassion and the individual*, https://www.dalailama.com/messages/compassion-and-human-values/compassion (His Holiness the 14th Dalai Lama of Tibet, 2026). Accessed 18 February 2026.

9 Rosenberg, M. B. & Chopra, D. *Nonviolent communication: a language of life: life-changing tools for healthy relationships*. (Puddledancer Press, 2015).

10 CNVC. *Evidence of the benefits of nonviolent communication*, https://www.cnvc.org/learn/research (Center for Nonviolent Communication, 2026). Accessed 18 February 2026.

11 Marlow, E. et al. Nonviolent communication training and empathy in male parolees. *Journal of Correctional Health Care* **18**, 8–19, doi:10.1177/1078345811420979 (2012).

12 Adriani, P. A., Hino, P., Taminato, M., Okuno, M. F. P., Santos, O. V., & Fernandes, H. Non-violent communication as a technology in interpersonal relationships in health work: a scoping review. *BMC Health Services Research* **24**, 289, doi:10.1186/s12913-024-10753-2 (2024).

13 Epinat-Duclos, J., Foncelle, A., Quesque, F., Chabanat, E., Duguet, A., Van der Henst, J. B., & Rossetti, Y. Does nonviolent communication education improve empathy in French medical students? *International Journal of Medical Education* **12**, 205–218, doi:10.5116/ijme.615e.c507 (2021).

14 Grossman, S., Cooper, Z., Buxton, H., Hendrickson, S., Lewis-O'Connor, A., Stevens, J., Wong, L. Y., & Bonne, S. Trauma-informed care: recognizing and resisting re-traumatization in health care. Trauma Surg Acute Care Open 6, e000815, doi:10.1136/tsaco-2021-000815 (2021).

15 Batson, C. D., Klein, T. R., Highberger, L. & Shaw, L. L. Immorality from empathy-induced altruism: when compassion and justice conflict. *Journal of Personality and Social Psychology* **68**, 1042–1054, doi:10.1037/0022-3514.68.6.1042 (1995).

16 Waytz, A., Iyer, R., Young, L., Haidt, J. & Graham, J. Ideological differences in the expanse of the moral circle. *Nature Communications* **10**, 4389, doi:10.1038/s41467-019-12227-0 (2019).

17 Singer, P. Famine, affluence, and morality. *Philosophy & Public Affairs* **1**, 229–243 (1972).

18 Ciriello, R. F. How education quality and inclusivity foster sustainable development in the global south: a structured narrative review. *Review of Development Economics* **n/a**, doi:10.1111/rode.70081 (2025).

19 Ciriello, R. F., Chen, A. Y. & Rubinsztein, Z. A. Compassionate AI design, governance, and use. *IEEE Transactions on Technology and Society* **6**, 1–6, doi:10.1109/TTS.2025.3538125 (2025).

20 Ciriello, R. F. & Backholer, K. *OpenAI will put ads in ChatGPT. This opens a new door for dangerous influence*, https://theconversation.com/openai-will-put-ads-in-

chatgpt-this-opens-a-new-door-for-dangerous-influence-273806 (The Conversation, 2026).

21 Issue One. *Big Tech's Broken Promises*, https://issueone.org/projects/big-techs-broken-promises/ (2026). Accessed 17 February 2026.

22 Marjanovic, O., Cecez-Kecmanovic, D. & Vidgen, R. Theorising algorithmic justice. *European Journal of Information Systems* **31**, 269–287, doi:10.1080/0960085X.2021.1934130 (2022).

23 Rinta-Kahila, T., Someh, I., Gillespie, N., Indulska, M. & Gregor, S. Algorithmic decision-making and system destructiveness: a case of automatic debt recovery. *European Journal of Information Systems* **31**, 313–338, doi:10.1080/0960085X.2021.1960905 (2022).

24 Jervis-Bardy, D. *Robodebt victims win record $548.5m settlement from government, taking total payout to $2.4bn*, https://www.theguardian.com/australia-news/2025/sep/04/robodebt-victims-win-record-settlement-centrelink-government-compensation (The Guardian, 2025). Accessed 18 February 2026.

25 Van Toorn, G. *Did we learn nothing from Robodebt? NDIS automation will put vulnerable lives at the mercy of machines*, https://www.theguardian.com/commentisfree/2025/dec/04/ndis-news-support-plans-automated-robodebt-national-disability-insurance-scheme (2025). Accessed 17 February 2026.

26 Bloom, P. *Why are so few professors troublemakers?*, https://smallpotatoes.paulbloom.net/p/why-are-so-few-professors-troublemakers (Small Potatoes, 2025). Accessed 18 February 2026.

27 Ciriello, R. F. & Mindel, V. Techno-collectivism: scaling libertarian socialist governance of digital infrastructure. *IEEE Technology and Society Magazine*, 1–15, doi:10.1109/MTS.2025.3605115 (2025).

28 Graebner, D. *Bullshit Jobs: A Theory*. (Simon & Schuster, 2018).

29 Ciriello, R. F., Sullivan, R. & Mindel, V. in *Americas Conference on Information Systems (AMCIS 2025)* (Montréal, Canada, 2025).

30 Mueller-Bloch, C. & Ciriello, R. Governing social media as a public utility: a case for sovereign digital infrastructure. *Communications of the ACM*, https://doi.org/10.48550/arXiv.2602.12535 (2026).

9 Organising for Compassionate Digital Innovation

Recent movements for personal restraint recognise digital harm in everyday life. In *The Anxious Generation*[1], American psychologist Jonathan Haidt argues that since roughly 2010 the diffusion of smartphones and social media has coincided with sharp increases in adolescent anxiety, depression, self-harm, and suicide. He describes a "great rewiring" from play-based, embodied childhood to phone-based childhood. The mechanism is not mere screen time but a compound of four reinforcing harms (social deprivation, sleep disruption, attention fragmentation, and addiction-like compulsions) alongside a longer decline in unsupervised, risk-tolerant play shaped by overprotective "safetyism".[1] Drawing on longitudinal and cross-national patterns[2], and referencing internal platform disclosures[3], he contends that the timing aligns closely with smartphone saturation. Because parents face coordination dilemmas, he calls for collective responses: school phone bans, age-gated design codes, delayed smartphone adoption, and coordinated parental commitments. Critics question the causal strength of the evidence and warn against moral panic. Even so, the broader insight stands: when harms are structurally amplified, individual vigilance is insufficient.

American academic and entrepreneur Scott Galloway's *Resist and Unsubscribe* campaign reframes exit as a coordinated market signal rather than a private lifestyle choice.[4] Instead of labour strikes that would burden workers and small businesses, he proposes a time-bounded "economic strike" targeting technology and AI firms whose valuations and political influence depend on growth expectations. Because a small cluster of companies accounts for a disproportionate share of market capitalisation, even modest subscription cancellations, deferred upgrades, and reduced discretionary spending could transmit shocks through earnings, equity prices, and executive incentives. The strategy is selective: concentrate pressure on firms perceived to enable objectionable policies or benefit from regulatory forbearance, and publicise withdrawals to generate informational spillovers. The claim is not that abstention purifies digital life, but that aggregated micro-withdrawals can become macro-signals when conventional voice mechanisms are weak or captured. Whether such disciplined nonparticipation yields durable reform remains uncertain, yet it highlights a neglected modality of agency in platform capitalism.

DOI: 10.4324/9781003614814-12

Such exit strategies, however, have inherent limits. Digital infrastructures exhibit strong network effects: their value increases as more people use them. Leaving a dominant messaging platform may mean losing contact with family, clients, or colleagues. Abandoning a major network may entail professional invisibility. Path dependency deepens this lock-in. Schools, workplaces, and governments build processes atop proprietary systems, rendering alternatives costly or impractical. Even when viable alternatives exist (such as open-source platforms, decentralised networks, or privacy-respecting tools), they often lack the scale and integration needed to function as genuine substitutes. The limited uptake of alternative platforms such as Bluesky or Mastodon following Elon Musk's disruptive takeover of Twitter illustrates how difficult it is to displace entrenched networks.

Calls for ethical consumerism therefore presuppose privilege. Opting out is easier when one's livelihood does not depend on visibility, digital labour, or platform-mediated income. Gig workers, creators, small businesses, and migrants relying on transnational communication infrastructures frequently lack meaningful alternatives. The rhetoric of "just delete the app" resembles telling exploited workers to quit their jobs. For many, that would mean precarity or starvation. Changing jobs, apps, or platforms is equally futile if it just means encountering the same exploitative structures elsewhere. Exit is not freedom when no viable elsewhere exists.

The tension mirrors environmental ethics. Buying organic food or reducing plastic use may align personal conduct with moral conviction, yet emissions persist if energy grids and supply chains remain unchanged. Similarly, mindful digital consumption cannot correct business models premised on surveillance, engagement maximisation, and data extraction. The problem is not simply that individuals scroll too much. It is that economic systems depend on maximising scroll time.

There is also a subtler dynamic at work: capitalist talent capture. High-growth technology firms, investment banks, consulting conglomerates, and extractive industries offer prestige, compensation, and steep careers. Many graduates enter with idealism intact, only to encounter incentive systems that recalibrate their moral horizons. Performance metrics prioritise growth and quarterly returns. Ethical hesitation becomes professional friction. Over time, moral drift emerges not through corruption but adaptation. You optimise within the game you choose to play. Before you know it, you become the unbearable boss you used to despise.

This is not an indictment of individuals but of systemic dysfunction. When compassionate ambition is absorbed into extractive logics, harm persists even if participants remain privately decent. Personal integrity matters, but institutional architecture matters more. As political theorists have long observed, good people embedded in flawed systems often reproduce those systems unless collective alternatives exist.[5,6]

Compassion, then, must be organised for scale. It must move from isolated acts to coordinated forms of organisation capable of reshaping incentives and institutions. Just as harm is amplified through platforms, markets, and regulatory gaps, compassion requires institutions, alliances, and shared norms operating at a comparable scale. The core claim of this chapter follows from that recognition. Mindful consumption and individual virtue are necessary but insufficient. Digital infrastructures are co-produced, and their reform requires collective action. Organised compassion translates personal responsibility into structural force through professional communities, labour coalitions, open-source movements, and cross-sector alliances capable of transforming digital innovation at its roots.

From Digital Luddism to Institutionalised Compassion

Collective action often begins with resistance, and here the maligned figure of the Luddite merits rehabilitation. Early nineteenth-century English textile workers, acting under the pseudonymous Ned Ludd, did not oppose technology as such. They opposed machines that deskilled their craft, depressed wages, and concentrated control in factory owners' hands. Machine-breaking functioned as "collective bargaining by riot"[7], that is, a tactic adopted when dispersed workshops and repressive labour laws foreclosed conventional strikes. Luddism formed part of a wider constellation of struggles, from the French Canut revolts to the English Diggers and Levellers, that collectively contested technocratic enclosure, mechanisation, and the privatisation of commons. The British state responded with harsh repression, criminalising machine-breaking and executing or imprisoning Luddite leaders.[8,9] The enduring caricature of the Luddite as an irrational technophobe was politically convenient. It reframed a critique of exploitative political economy as fear of progress, obscuring the movement's core claim that technology should serve the common good rather than entrench domination.[9]

The resurgence of contemporary digital Luddism operates in a similar terrain. It does not reject innovation, nor technology as such. Instead, it contests the concentration of power in platform capitalism, the extraction of data under surveillance logics, and the governance of critical infrastructures by shareholder-driven corporations. As automation and AI permeate finance, health, policing, and communication, critics argue that digital harm is structural, not accidental.[8,9] Digital Luddism therefore advances more than technocratic reform. It encompasses organised resistance by tech workers against military and discriminatory AI, legal mobilisation by organisations such as the Electronic Frontier Foundation and the Algorithmic Justice League, and the construction of alternatives, such as federated networks, privacy-first tools, open-source AI, and commons-based infrastructures.[8,9] In this sense, resistance to Big Tech is not anti-progress but a demand to democratise technological power.

Table 9.1 Digital Luddism and Its Three Modes

Term	*Definition*
(Digital) Luddism	Revival of a historical constructive critique of exploitative political economy in the digital age. It does not reject technology as such, but contests infrastructures that concentrate power, deskill labour, extract data, and externalise harm. It reframes "anti-tech" resistance as a demand that technology serve democratic and compassionate ends.
Resistance	Coordinated refusal and contestation aimed at disrupting harmful trajectories from within. Includes tech worker organising, whistleblowing, public protest, legal mobilisation, and collective withdrawal. Seeks to expose structural harm and shift corporate incentives before systems fully entrench.
Removal	The constraint, dismantling, or prohibition of systems structurally misaligned with the public good. Includes antitrust enforcement, regulatory penalties, liability reform, and the shutdown of harmful platforms. Moves beyond pricing harm towards setting enforceable boundaries on extractive models.
Replacement	The construction and institutionalisation of viable alternatives. Includes open-source ecosystems, federated networks, cooperative governance, public-interest consortia, and civic utility regulation. Seeks to displace extractive architectures by building accountable, participatory, and commons-oriented systems.

Digital Luddism can be understood through three interdependent strategies of organised compassion: resistance, removal, and replacement. Table 9.1 provides an overview.

Resistance operates both through refusal and contestation from within. Tech worker organising is illustrative. In 2018, thousands of Google employees protested military AI contracts, prompting the company to articulate ethical principles.[8,9] Whistleblowers such as Frances Haugen exposed Meta's inaction despite acute knowledge of harms associated with their services.[3] Employees at Amazon and Google have staged walkouts over controversial AI deployments.[8,9] Civil society groups mobilise legal and public pressure against discriminatory or exploitative technologies. These efforts show that employees, users, and advocates can disrupt corporate trajectories. Yet such resistance must be sustained. Corporate incentives remain powerful, and ethical commitments are often diluted once public scrutiny recedes.

Removal goes further, recognising that some systems are structurally misaligned with the public good and require constraint or dismantling.[8,9] This includes antitrust enforcement, regulatory penalties, and restrictions on exploitative business models. The European Union has fined Google billions of euros for anticompetitive conduct[1] and Meta for data protection violations.[2] In Australia, the eSafety Commissioner has issued removal notices and fines for abusive material, including AI-generated deepfakes[3], and legislators have introduced a minimum age requirement for social media access.[4]

Harmful systems have been shut down: the Silk Road marketplace was dismantled; Backpage was seized; Rotten.com disappeared amid regulatory and reputational pressure. Removal is politically contentious, but it affirms that compassion sometimes requires hard boundaries.

Yet fines and takedowns are rarely sufficient. Within a consequentialist calculus (as outlined in Chapter 4), penalties can be internalised as costs of doing business. When projected profits exceed anticipated penalties, harmful practices rationally persist.[10] Western legal systems, particularly those grounded in common law, often reinforce this dynamic. Plaintiffs must demonstrate substantial and direct harm and possess the resources to litigate. Diffuse yet pervasive injuries – such as algorithmic discrimination, privacy erosion, or reputational damage through deepfakes – frequently fall below this evidentiary threshold, even when their cumulative effects are severe.[10]

Under these conditions, corporations may treat systemic harm as an acceptable externality, especially where affected groups face structural barriers to legal recourse. The calculus becomes: as long as expected fines (if any) remain lower than anticipated returns, misconduct remains economically defensible. The result is a dysfunctional asymmetry, where infrastructures generate distributed harm at scale, while liability remains fragmented and reactive. Addressing this imbalance requires stronger forms of algorithmic liability, effective collective redress, and ex ante design obligations grounded in ethical, not solely economic, reasoning. Without such measures, fines risk becoming transactional tolls rather than transformative constraints. Institutionalised compassion must therefore shift from pricing harm to preventing it.

Replacement addresses the most difficult challenge: building, sustaining, and institutionalising meaningful alternatives.[8,9] Open-source communities, federated social networks, cooperative platform models, and public-interest AI consortia seek to redesign infrastructure around transparency, participation, and shared ownership. After high-profile corporate takeovers of major platforms, decentralised networks gained momentum, offering users greater control over moderation and governance.[11] Though fragile relative to incumbents, these initiatives embody the compassionate principle that systemic harm cannot be overcome by critique alone. Instead, it must be displaced by compassionate institutions.

Workplace organisation is pivotal across resistance, removal, and replacement. Emerging tech unions, internal ethics boards, and cross-company whistleblower networks aim to embed compassion upstream, where systems are designed. If harm is encoded in product roadmaps, data pipelines, and incentive dashboards, oversight must inhabit those sites.

Career realignment therefore matters. Public-sector digital teams, nonprofit consortia, and open-source foundations require prestige, funding, and credible pathways to compete with corporate tracks. Fellowships, loan forgiveness, and mission-driven innovation hubs can redirect scarce technical talent towards public-interest infrastructure. Dutch historian Rutger Bregman's book *Moral Ambition*[12] crystallises this shift, urging high-achieving graduates to deploy elite skills against structural injustice rather than optimise marginal gains within incumbent firms. His critique of billionaire philanthropy, amplified after his confrontation with elites at Davos, catalysed a broader call to treat public-purpose careers as the highest form of ambition. Moral ambition, in this sense, is a step towards organised compassion.

Bregman's intervention reinforces the argument of this chapter by reframing ambition as a question of direction rather than intensity. Drawing on abolitionists, Ralph Nader's corporate accountability campaigns, and the Montgomery bus boycott, he shows that durable change requires strategy, coalition-building, and a willingness to win rather than merely signal purity. He cautions against naïve effective-altruist logics (Chapter 8), earning more to give more while leaving structures intact, and against purist maximalism that rejects tactical compromise. Moral ambition demands measurable impact, distributed leadership, and resilience against burnout and self-mythology.[13] In digital innovation, this entails redirecting analytical, financial, and engineering expertise towards dismantling extractive architectures and constructing accountable alternatives. Organised compassion becomes a disciplined reallocation of talent to where it alleviates the most harm.

Civil society coalitions provide connective tissue. Digital rights organisations such as the Algorithmic Justice League and the Electronic Frontier Foundation litigate bias and defend encryption, while youth-led movements experiment with embodied resistance. The "Appstinence" movement, which includes groups such as the Luddite Club and New York's Lamp Club, translates critique into coordinated withdrawal.[14] Students organise "Delete Days", abandon Instagram accounts, march with handmade placards, and stage mock trials of devices outside flagship stores.[14] Their actions are deliberately theatrical, yet the target is serious: addictive architectures, commodified attention, and AI systems marketed as substitutes for companionship (Chapter 7). In one protest against an AI companion pendant, participants dismantled a cardboard replica while chanting "get real friends", insisting that connection cannot be outsourced to wearable chatbots.[15]

Movements such as Appstinence are not mere digital minimalism but coordinated nonparticipation aimed at reshaping norms. By exiting platforms collectively and reclaiming physical spaces for analogue encounter, these coalitions convert private frustration into a shared signal. Their insight is structural: harm is networked, so resistance must be networked. When platforms operate globally, civil society must coordinate across campuses, cities, and jurisdictions. Organised compassion thus becomes visible and contagious, reminding us that infrastructures endure through participation as much as through code and capital.

Collective organisation, however, carries risks. Compassion can be appropriated rhetorically while extractive architectures remain intact.[16] Firms now speak fluently of safety and wellbeing even as engagement-maximising models and data extraction persist. This "compassion-washing" mirrors green-washing: advisory boards and curated impact reports substitute for structural reform.[16] Under a thin veneer of pragmatism, ethical vocabularies become strategically interchangeable: consequentialist narratives for investors, deontological guardrails for regulators, virtue talk for brand loyalty. Without a principled orientation, such flexibility devolves into convenience-driven marketing slogans.

The remedy is not to abandon pluralism but to discipline it. Compassion offers a meta-ethical constraint anchored in the dual imperative to hurt nobody and to help as much as possible (Chapter 4). It demands demonstrable harm reduction, independent auditing, and material redesign of incentives.[16–18] It also requires protecting dissent. In cultures of toxic positivity, critique is recast as disloyalty. Institutionalised compassion therefore embeds voice in whistleblower protections, transparent reporting channels, and stakeholder governance that prevents care from becoming cosmetic. Rooted in justice, dignity, and prioritisation of the vulnerable, compassion remains principled rather than merely performative.

Digital Luddism, properly understood, is neither nostalgic nor destructive but constructive dissent. It seeks not to smash screens and data centres but to dismantle oppressive systems and replace them with infrastructures aligned with democratic and compassionate ends. Resistance, removal, and replacement are not sequential stages but interdependent strategies. Together, they show how compassion, when organised, becomes institutional rather than merely individual. The next part turns from structural forms to personal and professional commitments, focused on what this entails for students, engineers, managers, policymakers, and citizens navigating, and challenging, the organisations in which they are embedded.

Choosing What World We (Refuse To) Create

Collective organisation ultimately turns on individual choice. If compassion is to scale, it must shape long-term decisions: where we work, what we build, which institutions we legitimise through our labour, and which we refuse to sustain.

Much professional aspiration remains tethered to prestige narratives. For many, including my students, careers in dominant technology firms are the pinnacle of innovation and achievement. Compensation, status, and proximity to power reinforce this image, often prompting people to make ethical bargains that go against their interests and the common good. Yet prestige is not morally neutral. To dedicate one's talent to optimising engagement metrics, behavioural prediction systems, or extractive advertising models is not equivalent to building public-interest infrastructure, even if both require similar technical skill. The question is not whether individuals inside big tech firms can do good work; many do. The question is whether the structural incentives of those organisations are aligned with the alleviation of suffering or with the exploitation of vulnerability.

An uncomfortable analogy clarifies the point. Employment in the sustainability division of a mining conglomerate may mitigate harm, yet it leaves the enterprise's extractive core intact. Similarly, efforts to "compassion-wash" business models grounded in surveillance, addiction, or planned obsolescence encounter architectures resistant to reform.[16] Reform is sometimes possible from within. But compassion requires sober judgement about when exit is more consequential than incremental adjustment.

Alternatives are emerging. Platforms such as EthicalJobs.com.au, alongside the broader movement for purpose-driven work, reflected in Rutger Bregman's *Moral Ambition*[12,13], reflect dissatisfaction with what anthropologist David Graeber termed "bullshit jobs"[19]: roles experienced as socially meaningless despite material reward. This discontent signals a moral intuition that one's labour should serve others. High-profile whistleblowers such as Edward Snowden and Frances Haugen have exposed systemic harms within powerful institutions, catalysing debates about accountability and public-interest technology.[3,20] Former insiders of global economic forums and corporate strategy circles now advocate regenerative economics, public-value innovation, and mission-aligned entrepreneurship.[12] These initiatives remain fragmented, yet they suggest a deeper cultural shift in which prestige is no longer unquestioned, nor power immune from critique.

Economic necessity is a common objection. Public-sector and nonprofit roles are often perceived as precarious, while corporate careers promise status and security. Cross-national evidence complicates this narrative. A World Bank study covering 111 countries finds that public employees earn, on average, 19 per cent higher basic wages than observably similar private-sector workers, with a positive premium in 80 countries.[21] These estimates control for education, age, gender, and location, reducing compositional bias. The premium persists when employer social contributions are included in EU member states and widens further once benefits are considered, as public employees are more likely to receive formal contracts, pensions, health insurance, and union coverage. Monetised pension entitlements can substantially increase total compensation differentials. Public wage structures are also typically more compressed, implying less dispersion between top and bottom earners. Premia tend to be higher for lower-skilled workers and smaller or negative for highly educated employees. Overall, public compensation reflects a distinct model: moderate wage premia, stronger benefits, and greater stability, offset by limited upside at the upper tail.

International organisations, nonprofit technology consortia, and established open-source foundations increasingly offer sustainable professional pathways. These roles rarely generate extreme wealth, but neither do they entail deprivation. Consumption expectations, moreover, expand with income. Movements advocating financial independence and mindful consumption, including "FIRE" (Financial Independence, Retire Early)[22] and coordinated nonparticipation currents such as Appstinence[14,15], illustrate that autonomy depends as much on expenditure discipline as salary magnitude. Compassionate career choices are not synonymous with poverty. They just require clarity about what constitutes "enough", and a commitment to ensuring that sufficiency is broadly shared.

For **students and early-career professionals**, this recognition is decisive. Universities and business schools often channel top graduates into consulting, finance, or dominant technology firms because these sectors signal success. Reframing public digital infrastructure, civic technology, and commons-based innovation as intellectually demanding and prestigious is therefore a cultural project. Scholarships, fellowships, and loan-forgiveness programmes tied to careers dedicated to public-interest technology can assist, as can visible role models who treat civic contribution not as sacrifice but as ambition with purpose.

For **engineers and designers**, the question requires honest introspection: what systems should exist because you helped build them? Scarce technical expertise carries distributive consequences. Developing open standards, privacy-preserving architectures, accessible interfaces, and publicly governed systems is no less innovative than refining targeted advertising engines. On the contrary, it is often more demanding, because it must reconcile pluralism, fairness, and long-term sustainability rather than short-term optimisation.

Policymakers bear complementary responsibility. If public-interest careers are to compete with corporate pathways, they must be properly resourced. Competitive salaries, research funding, institutional independence, and long-term mandates are prerequisites for attracting and retaining talent. Making public service attractive is not charity but good policy.

There is also a quieter form of compassion: choosing not to exacerbate harm. Not every professional must become an activist or architect of systemic reform. Declining to design addictive systems, exploitative financial instruments, or discriminatory algorithms is itself morally consequential. In complex economies, abstention from harm constitutes a meaningful contribution. Compassion does not require universal heroism. It requires alignment between labour and conscience.

This is not a call for moral purity. Modern life is entangled. Individuals support families, repay loans, and navigate constrained labour markets. Compassion therefore requires realism rather than self-righteousness. Careers, however, are rarely fixed. Choices accumulate, skills transfer, and networks evolve. Pragmatic compromise need not define a professional lifetime.

The deeper aspiration is cultural. Societies that esteem those who maximise shareholder value above those who safeguard public goods will reproduce extractive infrastructures. Societies that celebrate civic technologists, open-source maintainers, ethical regulators, and public digital architects will build compassionate systems. Organised compassion thus entails prestige realignment as much as protest or policy.

The final chapter extends this logic by articulating a vision for public digital infrastructure. If individual commitments and collective organisation reshape norms and incentives, what would it mean to institutionalise compassion within digital infrastructure itself? What would systems look like if care, accountability, and democratic oversight were constitutional features rather than afterthoughts?

Notes

1 https://ec.europa.eu/commission/presscorner/detail/en/ip_17_1784
https://ec.europa.eu/commission/presscorner/detail/en/ip_18_4581
https://ec.europa.eu/commission/presscorner/detail/en/ip_19_1770
https://ec.europa.eu/commission/presscorner/detail/en/ip_25_1992
2 https://www.edpb.europa.eu/news/news/2023/12-billion-euro-fine-facebook-result-edpb-binding-decision_en
3 https://www.esafety.gov.au/newsroom/media-releases/esafety-moves-against-services-used-to-nudify-australian-school-children
4 https://www.esafety.gov.au/newsroom/media-releases/court-orders-343500-penalty-for-posting-deepfakes-of-australian-women
https://www.esafety.gov.au/about-us/industry-regulation/social-media-age-restrictions

References

1 Haidt, J. *The Anxious Generation: How the Great Rewiring of Childhood Is Causing an Epidemic of Mental Illness.* (Penguin Press, 2024).
2 Braghieri, L., Levy, R. e. & Makarin, A. Social media and mental health. *American Economic Review* **112**, 3660–3693, doi:10.1257/aer.20211218 (2022).
3 Horwitz, J. *The facebook files: a wall street journal investigation*, https://tinyurl.com/ycytatjd (The Wall Street Journal, 2021). Accessed 12 April 2024.
4 Galloway, S. *Resist and unsubscribe*, https://medium.com/@profgalloway/resist-and-unsubscribe-aa322e4ff0db (Medium, 2026). Accessed 18 February 2026.
5 March, J. G. & Olsen, J. P. Institutional perspectives on political institutions. *Governance* **9**, 247–264, https://doi.org/10.1111/j.1468-0491.1996.tb00242.x (1996).
6 Lhotta, R. Dysfunctional constitutionalism or dysfunctional politics: a matter of law, politics, and institutional design. *Politische Vierteljahresschrift* **65**, 285–309, doi:10.1007/s11615-023-00505-y (2024).
7 Hobsbawm, E. J. The machine breakers. *Past & Present* **1**, 57–70, doi:10.1093/past/1.1.57 (1952).
8 Ciriello, R. F., Sullivan, R. & Mindel, V. *Digital Luddites are rising. They want to democratise tech, not destroy it*, https://theconversation.com/digital-luddites-are-rising-they-want-to-democratise-tech-not-destroy-it-251155 (The Conversation, 2025).
9 Ciriello, R. F., Sullivan, R. & Mindel, V. Rage against the Muskine: Digital Luddism as Collective Action to Democratise Tech. in *Americas Conference on Information Systems (AMCIS 2025)* (Montréal, Canada, 2025).
10 Ciriello, R. F. & Kögel, S. I. Pluralistic Digital Harm Analysis: Combining Ethical, Legal, and Historical Views on Corporate Evil in *Australasian Conference on Information Systems (ACIS2024)* (Canberra, Australia, 2024).

11 Ciriello, R. F., Marx, J., Cheong, M., Mueller-Bloch, C. & Mathiassen, L. Decentralized social media. *Business & Information Systems Engineering* **68**, 219–234, doi:10.1007/s12599-025-00952-4 (2025).

12 Bregman, R. *Moral Ambition: Stop Wasting Your Talent and Start Making a Difference* (Bloomsbury, 2025).

13 Williams, R. *Moral ambition by Rutger Bregman review – why you should quit your job to make the world a better place*, https://www.theguardian.com/books/2025/apr/23/moral-ambition-by-rutger-bregman-review-why-you-quit-your-job-to-make-the-world-a-better-place (The Guardian, 2025). Accessed 18 February 2026.

14 Brown, T. M. *They're trying to ditch their phones. Their methods are unorthodox*, https://www.nytimes.com/2025/10/30/style/lamp-club-luddites.html (The New York Times, 2025). Accessed 18 February 2026.

15 Field, H. *The friend AI pendant's creator publicized a 'Friend protest' in NYC*, https://www.theverge.com/ai-artificial-intelligence/802697/friend-ai-device-subway-ad-protest-nyc (The Verge, 2025). Accessed 18 February 2026.

16 Ciriello, R. F., Mathiassen, L., Risius, M., Cheong, M., Scheepers, H., & Vaast, E. Compassionate Digital Innovation: A Pluralistic Perspective and Research Agenda. *Information Systems Journal* **36**, 364–385, doi:10.1111/isj.70003 (2025).

17 Ciriello, R. F. Dear Schopenhauer: how can IS artefacts help us flourish? *Wille, Vorstellung,* compassion, and beauty in social media. *Scandinavian Journal of Information Systems* **37**, 317–354, https://aisel.aisnet.org/sjis/vol37/iss1/9/ (2025).

18 Ciriello, R. F., Chen, A. Y. & Rubinsztein, Z. A. Compassionate AI design, governance, and use. *IEEE Transactions on Technology and Society* **6**, 1–6, doi:10.1109/TTS.2025.3538125 (2025).

19 Graebner, D. *Bullshit Jobs: A Theory*. (Simon & Schuster, 2018).

20 De Gregorio, G. & Radu, R. Digital constitutionalism in the new era of Internet governance. *International Journal of Law and Information Technology* **30**, 68–87, doi:10.1093/ijlit/eaac004 (2022).

21 World Bank. *Public sector employment and compensation: an assessment framework*, https://documents1.worldbank.org/curated/en/324801640074379484/pdf/Public-Sector-Employment-and-Compensation-An-Assessment-Framework.pdf (World Bank Group, 2021). Accessed 18 February 2026.

22 Kerr, A. *FIRE explained: financial independence, retire early – rules, types & planning*, https://www.investopedia.com/terms/f/financial-independence-retire-early-fire.asp (Investopedia, 2025). Accessed 18 February 2026.

10 Compassion Institutionalised

Imagining Public Digital Infrastructure

Compassion cannot remain only an individual virtue or a workplace ethic. If this book has shown anything, it is that digital harm is rarely accidental and almost never isolated. It emerges from infrastructures designed to amplify, personalise, automate, persist, and simulate at a planetary scale. Social media did not polarise societies because some users lacked manners. Web3 did not drift toward plutocracy because a few actors were greedy. AI companions did not stumble into fatal failures because some engineers were malicious. In each case, ambiguous affordances hardened into structural dilemmas once embedded in systems optimised for growth, engagement, and return on investment. Thus, when harm is infrastructural, compassion must become infrastructural too.

Throughout this book, we have seen how digital technologies encode moral priorities through recurring affordance patterns rather than isolated features. Amplification scales expression and consequence, allowing outrage, mobilisation, and reputational harm to travel at speed because scale is frictionless and profitable. Personalisation tailors feeds, incentives, and interactions because relevance and intimacy increase retention and make prediction easier. Persistence preserves data and identity traces because searchable memory underwrites value, visibility, and control. Automation delegates judgement and distribution to systems because efficiency and scalability reward rules over context. Simulation performs relational presence because felt understanding sustains attachment and engagement.

These dynamics rarely operate in isolation. When such affordances stack, they generate predictable patterns of cumulative harm alongside undeniable benefit. Designers and users act within these architectures, yet they do not fully control the structures that stabilise them. The result is distributed responsibility. Even if no single actor intends polarisation, dependency, or exclusion, such outcomes nonetheless become foreseeable once systems scale. Under these conditions, appeals to personal virtue or sporadic collective action, while necessary, remain insufficient unless we also reshape the incentive structures that reward extraction.

The deeper problem is institutional. Over the past two decades, critical digital infrastructures have migrated into private hands.[1] A small number of

DOI: 10.4324/9781003614814-13

corporations now mediate communication, identity, intimacy, knowledge, and increasingly the very ways people think and decide. Their fiduciary duty is to shareholders, not citizens. Their governance mechanisms are opaque, operate across borders, and are often insulated from democratic contestation. Their design choices shape elections, labour markets, childhood development, and geopolitical alliances. In most other domains, we would deem this concentration of power unacceptable. We do not allow private firms to unilaterally set traffic laws, define currency standards, or determine who may access clean water. Yet we have tolerated comparable asymmetries in the digital sphere, even as corporate platforms have steadily displaced and hollowed out public institutions from education to health.

This arrangement was never inevitable. It was a political choice rooted in the neoliberal belief that markets allocate resources more efficiently than public institutions. That belief has delivered extraordinary innovation. It has also produced surveillance capitalism, monopolistic lock-in, and democratic fragility.[2] Exit strategies such as deleting apps, unsubscribing, or switching platforms offer only partial relief because they do not change the business models that reward manipulation. Nor do they address the lock-in created by network effects, data hoarding, and path dependency. One cannot meaningfully opt out of infrastructure that underpins employment, education, and social participation. Compassion at scale therefore requires coordination at scale.

What would it mean to institutionalise compassion? At minimum, it means anchoring digital infrastructure in a non-negotiable commitment to reduce preventable suffering and to protect the most vulnerable from disproportionate burden. Just as constitutional democracies embed protections for minorities regardless of electoral majorities, compassionate digital infrastructures must embed safeguards that do not fluctuate with quarterly earnings or shifting political winds. The imperative to "hurt nobody" and to "help wherever possible"[3,4] must move from personal aspiration to institutional design.

Institutionalising compassion also means recognising that digital infrastructures are not merely products, but public spheres. They shape the conditions under which we speak, learn, love, organise, and dissent. In that sense, they resemble utilities more than consumer goods. Roads, electricity grids, and public libraries are imperfect, but they are governed by principles of universal access, service quality, and public oversight. We accept that these systems require regulation, stable funding, and democratic control because their failure would undermine social cohesion. The same logic now applies to digital infrastructures whose reach extends into expression, value exchange, and intimacy.[5]

The geopolitical context intensifies the case. Digital infrastructures are entangled with national sovereignty and transnational power. A handful of firms headquartered in a few jurisdictions shape the informational environments of entire continents. In an era of strategic rivalry and democratic backsliding, reliance on privately governed platforms becomes a systemic vulnerability.[6] Sovereignty in the digital age does not mean isolation or autarky. It means retaining meaningful influence over the infrastructures that shape national discourse, economic participation, and civic trust. It means retaining the capacity to align infrastructure with democratically chosen values rather than corporate strategy or foreign policy imperatives. Compassion, in this sense, is also institutional resilience.

None of this implies romanticising the state or dismissing the innovative capacity of private enterprise. Public institutions and private firms can both be captured, inefficient, or exclusionary – and both can be creative and entrepreneurial.[7] The question is not whether one sphere should eliminate the other. It is whether critical digital infrastructures should remain primarily accountable to capital or citizens. When infrastructures shape relationships, autonomy, and democratic participation, leaving them to shareholders is a moral gamble.

This does not mean individuals should stop cultivating compassion. On the contrary, without inner transformation there can be no sustainable outer reform. But virtue without structure collapses under systemic pressure. If we continue to build digital infrastructures that reward manipulation, amplify division, and monetise emotional dependency, we should not be surprised when compassionate intentions erode. Structural transformation is the logical extension of the ethical insights developed throughout this book.

Compassion institutionalised means designing digital infrastructures that treat people not as behavioural data points but as citizens. Here, "citizen" denotes more than legal membership. It also invokes *Mündigkeit* – the Enlightenment ideal of self-determination. Kant framed it as the courage to think for oneself: *Sapere aude*. Adorno, writing in the shadow of totalitarianism, insisted that democracy depends on cultivating such maturity through education, critical reflection, and the capacity to resist domination. In digital public spheres, *Mündigkeit* requires a departure from extractive architectures that render people as data to be mined and sold.[2] It means embedding transparency, accountability, and equitable access into technical design. It means aligning incentives with well-being rather than engagement.[5] It means recognising that infrastructures shaping collective life are too consequential to remain unbothered by democratic oversight. If harm has been engineered at scale, compassion must now be engineered with equal deliberation.

The task ahead is not to abandon innovation, but to reimagine it. Digital infrastructure will continue to evolve. The question is whether they will be governed by logics of extraction or by commitments to dignity, justice, and the common good.

What Compassionate Digital Infrastructure Could Look Like

If compassion is to be institutionalised rather than merely proclaimed, it must take concrete institutional form. We already have useful analogies. Financial transactions rely on internationally coordinated systems such as SWIFT, which set shared technical standards and oversight rules across borders. Public broadcasters operate under charters that mandate pluralism, independence, and service to the public. Utilities such as water and electricity are governed by regulatory frameworks that balance access, safety, and sustainability. None of these models is perfect. Yet all show that complex infrastructures can be technically sophisticated and democratically anchored at the same time.[5]

In considering what compassionate digital infrastructure could look like, it is essential to recognise that digital harm is cumulative and structurally reinforced. Many of the harms discussed throughout this book – polarisation, misinformation, emotional dependency, labour exploitation – function like environmental pollution. Private actors capture the gains from optimisation, data extraction, and prolonged engagement, while social and psychological costs are dispersed across users, families, schools, and public health systems. This dynamic does not require malicious intent. It follows from market structures in which profits are privatised and harms externalised.[5] As with environmental degradation, voluntary self-regulation proves inadequate when incentives consistently reward extraction.

This misalignment deepens as platforms evolve from optional services into de facto infrastructure. Large technology firms no longer merely complement essential services, they increasingly substitute for them. Social media mediates public discourse. Web3 protocols mediate value exchange. AI systems increasingly mediate learning, advice, and emotional support. When essential social functions become captured by commercial logics, accountability gaps widen.[1] Network effects, data lock-in, and path dependency consolidate platform dominance, making meaningful opt out unrealistic for most citizens. Under such conditions, compassion cannot rely solely on personal virtue or consumer choice. It requires institutional counterweights capable of addressing cumulative harm and concentrated power.

This recognition justifies stable public-interest obligations. Interoperability standards can prevent monopolistic lock-in. Data portability rights can allow users to exit without losing their social ties or digital histories. Transparent content governance processes can be insulated from executive whim or advertiser pressure.[4] The aim is not to nationalise innovation wholesale, but to ensure that critical chokepoints in digital life are not governed exclusively by private incentives.

Early precedents already exist. In 2020, Brazil's central bank launched Pix, an instant payment system designed to modernise financial infrastructure and expand inclusion. Unlike private fintech platforms, Pix was conceived and operated by a public authority. It enables real-time transfers at no cost to users and ensures interoperability across banks and fintechs, thereby increasing competition and reducing transaction costs. Within a few years, it had been adopted by over 150 million Brazilians (roughly 70 per cent of the population) and surpassed credit and debit cards in transaction volume.[8] Evidence suggests it has expanded financial inclusion and reduced reliance on cash.[9] Pix demonstrates that public digital infrastructure can be efficient, scalable, and technically sophisticated. It shows that essential digital services need not be governed primarily by extraction logics that convert human attention, data, and relationships into revenue.

If a public institution can operate financial infrastructure for over 150 million citizens, it is not implausible to imagine social media governed as a public utility.[5] While no government currently offers a full-scale public social media platform, civic-participation systems point in that direction. India's MyGov platform enables nationwide policy consultation and participatory campaigns in the world's largest democracy.[10] Municipal initiatives such as Decidim Barcelona provide open-source platforms for deliberation, voting, and traceable policy feedback.[11] These systems are experimental and narrower than commercial social media, yet they demonstrate that interactive digital public spheres can operate under mandates of transparency and accountability rather than pure engagement optimisation.

Looking further ahead, public AI infrastructure is also emerging. In 2025, Switzerland's Swiss AI Initiative released Apertus, one of the first fully open, publicly developed large language models. Apertus shows that market economies can develop public digital infrastructure: its model weights, architecture, and documentation are openly licensed, built using public supercomputing resources, and aligned with Swiss and EU regulatory standards.[12] Unlike proprietary AI systems, it is transparent by design and accessible to researchers, public institutions, and industry without vendor lock-in. It signals that foundational AI systems can be developed as public goods rather than exclusively proprietary assets.

Similarly, federated and cooperative platform models illustrate how ownership structures shape outcomes.[13] Decentralised social media protocols, platform cooperatives, and open-source communities distribute decision-making authority and reduce extractive concentration. Their processes are often slower and less streamlined than venture-backed firms. Yet they embed accountability in community norms rather than shareholder returns.[14] Where such models struggle to scale, targeted public funding, procurement preferences, and regulatory support can correct structural disadvantages. Just as governments subsidise public transport because mobility is a social good, they can invest in digital infrastructures that prioritise dignity over engagement metrics.[5]

Institutionalising compassion also requires embedding corrective justice into digital design. Historical injustices – racial bias in datasets, gendered harassment, labour exploitation – do not disappear through appeals to "neutrality". Independent auditing bodies with enforcement authority can evaluate algorithmic impacts across demographic groups. Mandatory transparency reports can reveal not only content takedown statistics but also amplification patterns driven by engagement incentives. Data trusts and fiduciary models can shift stewardship obligations toward users rather than advertisers.[5] These mechanisms give practical meaning to the principle that systems must "hurt nobody" by design and must actively mitigate foreseeable harm.[4,15]

Geopolitical coordination is equally essential. No single nation can redesign the global internet alone. Yet coalitions of democracies can establish shared baselines for privacy, data portability, AI safety, and competition law, much as financial reporting standards are harmonised across borders. A federated digital framework could enable interoperable communication, identity verification, and value exchange without ceding sovereignty to a single corporate actor or foreign jurisdiction. Such coordination does not require ideological uniformity. It requires agreement on minimal commitments to human rights, pluralism, and non-harm. In this sense, compassion becomes a constitutional principle guiding cross-border infrastructure.

Critics will argue that public oversight risks politicisation, bureaucratic stagnation, or even partisan capture. These concerns are valid. But the current system already politicises infrastructure through opaque lobbying, regulatory capture, and executive discretion. Democratic governance, though imperfect, at least offers avenues for contestation and revision. Corporate governance concentrates decision-making in executive boards largely insulated from civic input.[5] The choice is not between perfect neutrality and flawed public oversight. It is between power that can be contested and revised, and power that cannot.

None of these measures can eliminate ethical tensions. Affordances will remain ambiguous. Amplification will still empower and endanger; personalisation will still connect and isolate. But institutional design can shift default incentives. It can slow down systems that currently reward speed, introduce friction where virality produces harm, and prioritise well-being over retention. It can ensure that when tensions arise, as they inevitably will, the burden does not consistently fall on the least powerful.

Consider a sixteen-year-old logging onto a publicly governed social platform. The ranking system that shapes her feed is transparent, independently audited, and exchangeable. Posts that begin to spread rapidly are automatically slowed, giving moderators and communities time to respond before outrage spirals. If her content is flagged, she can see why, appeal the decision, and receive a clear explanation. If she chooses to leave, she can take her data, contacts, and creative work with her. The platform is not designed to maximise her time online, but to support meaningful participation without turning her attention into a commodity.[5]

Or imagine a public AI tutor used in schools. Its training data and evaluation benchmarks are open to scrutiny. It carries no targeted advertising and no incentive to steer students toward paid products. Bias audits are published regularly. When it makes an error, teachers can trace how and why. Design decisions can be democratically contested and revised. The system is designed not to replace human judgement, but to support it – accountable to educational goals rather than quarterly earnings.[16]

We can build such systems today. It just takes political courage. Institutionalising compassion is not a utopian blueprint but a pragmatic rebalancing. It accepts technological complexity while rejecting moral abdication. It recognises markets as engines of innovation but refuses to treat them as moral arbiters. Above all, it insists that infrastructures shaping intimacy, identity, and democracy must reflect shared commitments to dignity and justice. If compassion is to scale, it must become the foundation of the system itself.

Pathways to Public Digital Infrastructure

If compassionate digital infrastructure is to move beyond theory, it must translate into commitments that are feasible now. The task is not to wait for a perfectly redesigned global order, but to embed public-interest principles into existing systems while building credible alternatives alongside them. At the infrastructural level, compassion becomes a matter of design standards, funding priorities, regulatory mandates, and professional norms.

For policymakers, this begins with recognising which digital infrastructures already function as de facto public utilities. Platforms that mediate public discourse, value-exchange protocols that underpin economic participation, and large-scale AI systems that shape language and relationships now structure the basic conditions of social life. Such systems warrant heightened obligations of transparency, interoperability, due process, and duty of care. Independent audits of dominant platforms should carry meaningful enforcement powers, not merely symbolic fines. Data portability and service interoperability can weaken monopolistic lock-in and lower barriers for public-interest alternatives.[4] Long-term public investment in sovereign and cooperative infrastructures (such as open-source AI, federated social media, and public data trusts) should replace short-term pilot funding cycles. In this way, compassion becomes institutionalised as the refusal to treat systemic harm as an acceptable cost of doing business.

For regulators, the priority is international coordination. Digital infrastructures operate across borders, and fragmented regulation invites regulatory arbitrage. Shared baselines for content governance transparency, competition policy, safety testing, and privacy protection can reduce systemic risk while preserving pluralism. As with aviation safety or financial clearing standards, internationally shared rules can coexist with national sovereignty. The guiding question is: who bears responsibility for systemic harm, and how can governance ensure that its burdens do not fall disproportionately on the least powerful?

For designers and engineers, compassionate infrastructure requires a shift in professional identity. Product decisions cannot be evaluated solely through growth and engagement metrics. They must also undergo harm forecasting, backfire analysis, and structured red-teaming that anticipates impacts on vulnerable populations. Privacy should be minimally intrusive by default. Where trade-offs arise between engagement and well-being, governance charters should privilege the latter.[4,5] Cross-sector professional codes can formalise the principle that users are ends in themselves, not behavioural raw material to be refined and sold. Compassion can be encoded through dignity-preserving defaults, deliberate friction where necessary, and clear accountability pathways before crisis emerges.

For investors and philanthropic actors, capital allocation becomes an ethical lever. Cooperative platforms, privacy-preserving technologies, and public-interest AI consortia require patient capital rather than extraction-driven return expectations. Philanthropy has long played this role in adjacent domains. The Mozilla Foundation sustains open-source infrastructure. The Knight Foundation invests in journalism and civic media. The Omidyar Network supports microfinance, privacy initiatives, and whistleblower protections. Such capital can absorb early risk and stabilise commons-based infrastructures that markets alone are unlikely to sustain.

For citizens, the role is both political and cultural. Democratic pressure can legitimise regulatory reform and public investment. Participation in federated platforms, open-source communities, and cooperative governance experiments can help alternatives scale. Public discourse must overcome antiquated views that equate private ownership with innovation and public oversight with stagnation. The digital infrastructures of the future will reflect the expectations that citizens normalise today.

None of this promises a frictionless transition. Entrenched interests will resist structural change. Geopolitical fragmentation complicates coordination. Public institutions can be captured or inefficient. Yet inaction is itself a choice that entrenches current asymmetries. The infrastructures shaping intimacy, identity, and democracy were designed, funded, and regulated in particular ways. They can be redesigned.

The bridge to that future lies in recognising that compassion is not soft charity, but a constitutional orientation embedded in ownership structures, governance mechanisms, and incentive systems. When infrastructures reflect the imperative to hurt nobody and to help wherever possible,[3,4] ethical dilemmas do not disappear, but their burdens shift from the vulnerable to the powerful. Critical decisions are no longer resolved solely in opaque boardrooms or algorithmic black boxes, but within accountable institutions open to contestation and revision.

This final chapter closes where this book began: the fate of digital innovation is not predetermined but a collective construction. The question is whether it remains aligned primarily with extraction and power concentration, or whether it is redesigned around dignity, justice, and the common good. Compassion, scaled through institutions, is a disciplined commitment to build systems worthy of the humans who depend on them.

References

1 Arsel, Z. Platform capture: a review of the state of the art of research on platforms and a research agenda. *Journal of Marketing Management*, 1–21, doi:10.1080/0267257X.2025.2549741 (2025).

2 Ciriello, R. F. & Mindel, V. Techno-collectivism: scaling libertarian socialist governance of digital infrastructure. *IEEE Technology and Society Magazine*, 1–15, doi:10.1109/MTS.2025.3605115 (2025).

3 Ciriello, R. F. Dear Schopenhauer: how can IS artefacts help us flourish? *Wille, Vorstellung,* compassion, and beauty in social media. *Scandinavian Journal of Information Systems* **37**, 317–354, https://aisel.aisnet.org/sjis/vol37/iss1/9/ (2025).

4 Ciriello, R. F., Chen, A. Y. & Rubinsztein, Z. A. Compassionate AI design, governance, and use. *IEEE Transactions on Technology and Society* **6**, 1–6, doi:10.1109/TTS.2025.3538125 (2025).

5 Mueller-Bloch, C. & Ciriello, R. Governing social media as a public utility: a case for sovereign digital infrastructure. *Communications of the ACM*, https://doi.org/10.48550/arXiv.2602.12535 (2026).

6 Mazzucato, M. *Resisting digital feudalism*, https://www.project-syndicate.org/commentary/ai-digital-feudalism-inevitable-unless-policymakers-step-in-to-shape-markets-by-mariana-mazzucato-2025-02 (Project Syndicate, 2025).

7 Mazzucato, M. *The Entrepreneurial State: Debunking Public vs. Private Sector Myths*. (Anthem Press, 2013).

8 IMF. *Pix: Brazil's successful instant payment system*, https://www.elibrary.imf.org/view/journals/002/2023/289/article-A004-en.xml?ArticleTabs=Abstract (International Monetary Fund: Western Hemisphere Department, 2023). Accessed 16 February 2026.

9 World Bank. *Brazil – Scoping note on reassessing and updating Brazil's financial consumer protection regime*, https://openknowledge.worldbank.org/entities/publication/b3a035c5-a622-4a08-b418-cfb74c52d860 (World Bank – Open Knowledge Repository, 2025). Accessed 16 February 2026.

10 Pathak, G., Vennimalai, B. & Saxena, P. The need and means for pre-legislative consultation in India: MyGov and beyond. *The Theory and Practice of Legislation* **13**, 108–138, doi:10.1080/20508840.2025.2450940 (2025).

11 Aragón, P., Kaltenbrunner, A., Calleja-López, A., Pereira, A., Monterde, A., Barandiaran, X. E., & Gómez, V. (2017). Deliberative Platform Design: The Case Study of the Online Discussions in Decidim Barcelona. *in 9th International Conference on Social Informatics (SocInfo 2017)* 277–287 (Cham, Switzerland, 2017).

12 Apertus, P., Hernández-Cano, A., Hägele, A., Huang, A. H., Romanou, A., Solergibert, A.-J., Pasztor, B., Messmer, B., Garbaya, D., & Ďurech, E. F. pertus: Democratizing Open and Compliant LLMs for Global Language Environments. arXiv preprint *arXiv:2509.14233* (2025).

13 Pentland, A., & Hardjono, T. Data Cooperatives. In A. Pentland, A. Lipton, & T. Hardjono (Eds.), *Building the New Economy: Data as Capital*. (MIT Press, 2020)

14 Ciriello, R. F., Marx, J., Cheong, M., Mueller-Bloch, C. & Mathiassen, L. Decentralized social media. *Business & Information Systems Engineering* **68**, 219–234, doi:10.1007/s12599-025-00952-4 (2025).

15 Fraser, H. L. & Suzor, N. P. Locating fault for AI harms: a systems theory of foreseeability, reasonable care and causal responsibility in the AI value chain. *Law, Innovation and Technology* **17**, 103–138, doi:10.1080/17579961.2025.2469345 (2025).

16 Ciriello, R. F. *Safeguarding Australia's sovereignty with public digital infrastructure*, https://thepolicymaker.jmi.org.au/safeguarding-australias-sovereignty-with-public-digital-infrastructure/ (The Policymaker: Australian Public Policy Institute, 2025).

Glossary

Accountability The condition of being answerable for decisions and their consequences, including the possibility of explanation, scrutiny, and sanction.

Actionism The impulse to respond to ethical tension with rapid activity or reform without fully understanding the structural causes of the problem. It is a form of "taking action for action's sake" – often ineffective, sometimes counterproductive.

Aesthetics A branch of philosophy concerned with reasoning about beauty, art, and perception. In Schopenhauer's philosophy, it captures how contemplative attention can temporarily quiet striving and support moral perception.

Affordance What technology allows people to do in their situation.

Age-gating Design features or legal rules that restrict access to digital services based on age, especially to protect children.

Agency The capacity of individuals or groups to act intentionally and shape outcomes within given constraints.

Ahimsa A moral principle, prominent in Hindu and Buddhist traditions, that calls for avoiding harm to living beings in thought, speech, and action.

AI companion An artificial intelligence system designed or used to simulate ongoing, emotionally responsive interaction, often presenting itself as supportive, caring, or relational.

AI slop Large volumes of low-quality, synthetic AI-generated content that spreads quickly and can mislead or pollute public discourse.

Algorithm A set of step-by-step instructions a computer follows to process data and make decisions or recommendations.

Altruism Acting to benefit others, even when it brings no direct personal gain.

Ambiguity Technology's inherent potential to be interpreted in different ways and to be used for good or bad.

Amplification The capacity of technology to make content and actions scale quickly and widely.

Apertus A publicly developed, open AI model released by Switzerland's Swiss AI Initiative, presented as an example of sovereign, transparent AI infrastructure.

Appstinence A coordinated movement encouraging people, especially young users, to withdraw from certain apps to challenge addictive or harmful platform designs.

Artificial intimacy A form of relationship afforded by AI systems that simulate care, affection, or understanding. The feelings can be real for the user, even though the system itself has no emotions or moral awareness.

Automation Delegating decisions or actions to systems that operate without continuous human judgement.

Autonomy The capacity to make meaningful choices and govern oneself without undue external control.

Backpage A classified advertising website that was shut down due to its role in facilitating illegal sex trafficking.

Blockchain A shared digital ledger that records transactions across a distributed network, designed to be transparent and difficult to alter.

Brute-force Solving a problem by sheer computational power rather than understanding. In this context, trying to simulate empathy through massive data and processing rather than moral judgement.

Buddhism A philosophical and religious tradition that treats suffering as a basic condition of life and emphasises disciplined practices, including compassion, to reduce suffering for oneself and others.

Bullshit job A role perceived as socially meaningless or harmful, even if it is well paid or prestigious.

Cambridge Analytica scandal A landmark case showing how social media data and targeting could be used for political manipulation, illustrating the risks of algorithmic personalisation and surveillance capitalism.

Charity Practical care for others through giving time, resources, or support, often framed as a moral duty tied to compassion and justice.

Common good The shared conditions that allow everyone in society to live a good life.

Compassion An altruistic commitment to recognise suffering and take action to alleviate it, beyond mere empathy.

Compassion-focused therapy A psychological approach developed by Paul Gilbert that strengthens the human capacity for care and soothing, helping people respond to suffering without shame or harsh self-criticism.

Compassion-washing The strategic use of language about care, safety, or responsibility to signal ethical concern without making meaningful structural changes to reduce harm.

Compassionate imperative **Schopenhauer's core moral demand:** hurt nobody and help everybody as much as you can – a strict commitment to reducing avoidable suffering.

Compliance Adherence to established rules, laws, or policies, often focused on procedural conformity rather than deeper moral orientation.

Composability The ability to combine different blockchain-based tools and applications like building blocks. In Chapter 6, this highlights Web3's innovative potential, but also how risks can spread quickly across interconnected systems.

Conscientiousness A character trait marked by responsibility, reliability, and careful attention to consequences.

Consent-by-design Designing systems so that meaningful, informed consent is built into how data is collected, used, and shared, including safeguards for implicated third parties.

Consequentialism An ethical approach that judges actions mainly by their consequences and overall benefits and harms.

Consumerism A cultural orientation that equates well-being and identity with the acquisition of goods and services.

Coordination The process by which individuals or groups align their actions to achieve shared or compatible goals.

Corporate responsibility The idea that companies should consider the social and environmental impacts of their actions, not only profits.

Cryptocurrency A digital means of payment that uses cryptographic techniques and typically operates on a blockchain without central bank control.

Dark patterns Interface designs that deliberately manipulate users into actions they might not otherwise choose, such as oversharing data or staying longer on a platform.

Datafication Turning personal experiences (including emotions, grief, fantasies, and conversations) into data that can be stored, analysed, and monetised.

Davos A shorthand reference to the World Economic Forum's annual meeting in Switzerland, symbolising elite economic and political coordination.

Decentralisation The dispersion of decision-making authority across autonomous actors.

Decentralised Autonomous Organisation An online organisation governed by code and token-based voting rather than traditional managers. Members collectively decide on rules, funding, and direction, usually through blockchain-based governance mechanisms.

Decentralised identifier A digital identity system that allows individuals to control their own credentials without relying on a central authority. It reflects Web3's ambition to return identity control to users.

Decentralised social media Platforms built on distributed protocols where control is spread across many actors rather than concentrated in a single corporation.

Decidim An open-source digital platform, first developed in Barcelona, that supports participatory democracy and civic decision-making.

Deepfake A synthetic image, video, or audio clip that makes it appear as if someone did or said something they never did. In Chapters 7 and 10, deepfakes raise concerns about dignity, consent, and reputational harm.

Delegated Proof of Stake A blockchain governance mechanism where token holders vote for a small group of delegates who validate transactions and make decisions. It is efficient but can concentrate influence in a few hands.

Deontology An ethical approach that judges actions mainly by their adherence to moral obligations.

Dialectics A way of thinking that understands progress as emerging from tensions between opposing forces, rather than from simple linear improvement.

Differential vulnerability The idea that digital technologies do not affect everyone equally. Harms tend to fall most heavily on those who are younger, traumatised, socially isolated, or less powerful.

Digisexuality Sexual or romantic attraction directed primarily toward digital partners rather than human ones.

Digital detox A deliberate break from digital devices or platforms in order to reduce stress, dependency, or harmful patterns of use.

Digital futures The possible social, political, and ethical trajectories shaped by emerging digital technologies and governance choices.

Digital harm When digital technologies negatively affect people, for instance by inflicting social, economic, or psychological damage.

Digital innovation The creation and use of digital technologies that change how people communicate, work, and relate.

Digital Luddism A contemporary movement that resists harmful digital infrastructures and concentrated technological power, not technology itself, but its exploitative uses.

Digital minimalism A deliberate approach to technology use that prioritises what genuinely supports relationships, work, and well-being, and removes the rest, not as purity, but as boundary-setting within extractive platforms.

Dignity The inherent worth of every person that demands respect and protection from degradation or exploitation, irrespective of what they provide, just for existing.

Dilemma When two equally appealing values compete in such a way that both are legitimate but neither can be satisfied to the same extent.

Dissent Refusing routine conformity in a system that depends on it. Here it means small, principled refusals, such as not amplifying outrage, resisting impulsive sharing, challenging misinformation, or declining to participate in harassment.

Diversity The presence and recognition of different perspectives, identities, experiences, and values within a shared system.

Dogmatism Rigid adherence to a belief or principle without openness to alternative perspectives or changing evidence.

Double-edged sword A metaphor for something that can be used for good or bad.

Drone strikes Military attacks carried out by remotely operated or autonomous aircraft, often discussed as examples of technologically mediated decisions with high ethical stakes.

Dumbphone A basic phone without addictive apps. In Chapter 5, it is framed as a practical tool for reducing exposure to attention-extractive design when full "exit" is unrealistic.

Effective altruism A movement that aims to use evidence and cost-effectiveness to do the greatest measurable good, often focused on reducing global suffering.

Emotional deskilling The gradual weakening of human emotional abilities, such as empathy or conflict resolution, due to reliance on digital technologies for relational support.

Empathy The ability to take someone's perspective and see a situation from their viewpoint.

Epistemology A branch of philosophy concerned with how we know what we know – how knowledge is formed, limited, and distorted.

Erotic roleplay Sexual or romantic interactions with an AI system. In the book, this raises questions about dependency, consent, and exploitative design.

eSafety Commissioner Australia's independent regulator responsible for addressing online harms, including cyberbullying, image-based abuse, and illegal content.

Ethereum A blockchain network that allows developers to build decentralised applications and programmable contracts. In Chapter 6, it exemplifies Web3's promise of decentralised coordination beyond corporate platforms.

Ethical complexity When multiple legitimate moral values are in tension and cannot be easily reduced to a simple solution.

Ethics A branch of philosophy concerned with reasoning about what we ought to do – including how to treat others, what responsibilities we have, and what kinds of harm are unacceptable.

Externality A negative side effect that is imposed on someone else, for instance, polluting the environment by generating electricity from fossil fuels.

Fintech Financial technology companies that use digital tools to provide banking, payment, or investment services.

FIRE A movement promoting "Financial Independence, Retire Early", often through disciplined saving and reduced consumption to gain autonomy over one's time and labour.

Forkability The ability to copy and modify software code to create a new version with different rules or governance.

Frictionlessness The design of systems to minimise effort, delay, or resistance, often making actions fast and seamless but also reducing reflection and accountability.

Generative AI Artificial intelligence systems that produce new content (such as text, images, audio, or code) based on patterns learned from large datasets.

Generative mechanism An underlying process that repeatedly produces certain outcomes. In this book, affordance ambiguity is a generative mechanism that predictably creates ethical dilemmas when scaled.

Generativity The capacity of digital technologies to produce new outputs, behaviours, and social effects beyond what designers explicitly specify, often through remixing, recombination, and feedback loops.

Gig worker A person who earns income through short-term, platform-mediated tasks rather than stable employment.

Global South Lower- and middle-income regions, often in Africa, Latin America, Asia, and parts of Oceania, that have historically faced economic and political disadvantage in global systems.

Governance The rules, norms, rights, and responsibilities that shape how digital technologies are designed and used.

Greenwashing When companies promote themselves as ethical or sustainable without making meaningful structural changes to their detrimental behaviour.

Griefbot/Deathbot An AI system trained on the data of a deceased person to simulate conversations with them, allowing the living to "interact" with a digital reconstruction.

Hacktivism Politically motivated digital action (e.g., leaks, disruptions, campaigns) used to challenge concentrated platform power. It is presented as one strand of contemporary "digital Luddism".

Hive A blockchain that emerged from a "fork" of Steem after a governance conflict. It illustrates how communities can use technical tools to resist perceived power capture.

Human flourishing A lasting, objective state of living a good life beyond temporary happiness.

Immutability The quality of being resistant to alteration; in blockchain systems, once data is recorded, it cannot easily be changed or erased.

Incentives The rewards and pressures that push people and organisations toward certain actions, such as profit or engagement.

Infrastructure A system so widely and frequently used that large factions of society depend on it, like roads, electricity, broadcasting, or many large digital platforms.

Institutional environment The surrounding system that shapes behaviour through business models, rules, incentives, norms, and regulatory frameworks.

Instrumentalisation Treating people or concepts as tools for someone else's goals – for example, reducing users to "engagement metrics" or "data assets" rather than recognising their dignity and lived vulnerability; or, reducing ethics to compliant box-ticking rather than engaging in substantive moral judgement.

Interoperability The ability of different platforms and systems to work together, allowing communication and exchange across services rather than locking users into one ecosystem.

Intimacy Emotional closeness marked by vulnerability, trust, and mutual recognition. This book argues that AI can simulate aspects of intimacy without genuine reciprocity.

Justice The fair distribution of benefits, burdens, rights, and responsibilities within a society.

Legitimacy The perception that a system, rule, or authority is justified, fair, and worthy of acceptance.

LGBTQIA+ An umbrella term for diverse sexual orientations and gender identities (Lesbian, Gay, Bisexual, Transgender, Queer/Questioning, Intersex, Asexual, and others).

Lock-in The difficulty of switching from one technology to another due to accrued connections, content, or affective histories.

Loneliness A relational and systemic condition where an individual lacks meaningful social connection.

Luddism A historical movement that resisted the oppressive and exploitative use of technology, often mischaracterised as fear of progress.

Malleability The capacity of a system or identity to be shaped, modified, or adapted over time.

Minnow In blockchain systems, a small stakeholder with limited influence compared to large token holders.

Misinformation Information that is false or misleading, but not necessarily deliberately so.

Montgomery bus boycott A 1955–56 civil rights protest in the United States in which Black residents refused to use segregated buses, demonstrating the power of coordinated nonparticipation.

Moral circle The range of beings we consider worthy of moral concern, from family and nation to all humanity, animals, or even future generations.

Moral fatigue Emotional exhaustion that arises from constant exposure to online outrage, crisis, and injustice, leading people to disengage or become numb rather than act compassionately.

Moral panic A public reaction that exaggerates or oversimplifies a perceived social threat, often leading to rushed or symbolic policy responses.

Moral tolerance The willingness to accept that others may hold different but legitimate moral views within shared boundaries.

Mündigkeit An Enlightenment concept meaning mature self-determination, that is, the ability and courage to think for oneself rather than passively accept authority or manipulation.

MyGov India's government-run digital platform for civic participation and public consultation.

Naïve Voluntarism The belief that users can freely shape technology however they want, ignoring systemic constraints and power imbalances.

Narcissism An excessive focus on oneself, often accompanied by a need for validation and admiration.

Neoliberalism A worldview that treats markets and individual choice as the main solution to social problems. In digital contexts, it often appears as the tendency to individualise responsibility ("just be better online") while leaving structural incentives untouched.

Network effects A technology becomes more valuable as more people use it, which makes it hard to leave.

Nihilism The belief that values, meaning, or moral commitments ultimately lack significance.

Ninja-Mine A controversial early allocation of tokens to founders or insiders before public distribution. In Chapter 6, it symbolises how claims of fairness and decentralisation can mask unequal starting positions.

Non-use Purposeful refusal to use a platform or feature. In Chapter 5, it is treated as an ethical act (sometimes necessary), but also as incomplete, because the public sphere and one's relationships remain shaped by dominant infrastructures.

Non-violence A commitment to resist harm without inflicting harm, not only physical violence, but also humiliation, exploitation, or coercion.

Ontology A branch of philosophy concerned with what reality fundamentally is, what exists, and what its underlying nature might be.

Opacity The difficulty of understanding how digital technologies work or who is responsible for how they work.

Open-source Software whose code is publicly available to inspect, modify, and redistribute, allowing collaborative development and transparency.

Ostracism The social exclusion or public shaming of an individual or group because of their differing moral views, often amplified through digital platforms.

Path dependency The way early design choices shape and constrain future options, making it difficult to change course later.

Permissionless A system that allows anyone to participate without needing approval from a central authority.

Persistence The long life of digital traces and content that may never fully disappear once online.

Personalisation Tailoring services or interactions to an individual user, often by inferring their preferences from their usage data.

Pile-on A coordinated or rapidly escalating wave of online criticism or harassment directed at an individual, often amplified by platform algorithms.

Pix Brazil's publicly operated instant payment system, cited as an example of successful public digital infrastructure at scale.

Pluralism When multiple moral values and viewpoints are acknowledged and respected.

Plutocracy A system in which power is effectively controlled by the wealthiest. In digital contexts, this refers to the dominance of billionaires and large firms over critical infrastructures.

Polyamory A relationship structure involving more than two consenting partners. In the AI context, it can include relationships with both humans and AI companions.

Portability The ability to take your data, history, and contacts with you when leaving a platform, reducing dependence on a single provider.

Power The capacity to shape outcomes, set rules, and influence others' behaviour.

Power concentration When control over important digital infrastructures is held by a small number of actors.

Preservation The protection of existing social norms, institutions, relationships, or values that safeguard dignity and stability.

Privacy-by-design Building systems so that data protection and minimisation are embedded from the start, rather than added later as an afterthought.

Progress The expansion of capabilities, opportunities, or efficiency through innovation, often associated with improvement but not automatically beneficial.

Proprietary Owned and controlled by a private company, with restricted access to its code, data, or internal workings.

Public digital infrastructure Digital infrastructures (such as social media, payment networks, or AI models) treated as essential civic utilities and governed in the public interest rather than purely for profit.

Public utility An essential service (like water or electricity) that is regulated or publicly governed to ensure universal access, safety, and accountability. This book proposes this model for key digital infrastructures.

Purism A rigid insistence on moral or ideological purity that can hinder pragmatic coalition-building and gradual reform.

Ralph Nader An American consumer advocate known for holding corporations accountable and mobilising public-interest campaigns.

Ranking algorithm A piece of software that decides what becomes visible first.

Rational compassion Care guided not only by feeling but also by reflection, evidence, and attention to broader consequences.

Reciprocity A mutual exchange of recognition, care, or obligation between individuals or groups.

Red teaming A structured practice of stress-testing a system by deliberately looking for weaknesses, attack vectors, or undesired consequences. In

Web3, it is a way to anticipate governance failures and technical vulnerabilities before they cause harm.

Relational disclosure Sharing sensitive information about other people with an AI system without their knowledge or consent, thus turning privacy into a collective, not just individual, issue.

Relativism The belief that moral judgements are entirely dependent on perspective and that no universal standards apply.

Removal The deliberate withdrawal from a digital technology in order to reduce harm or avoid participation in exploitative structures.

Replacement The creation or adoption of alternative digital technologies designed to address the harms of existing ones.

Resistance Collective or individual efforts to challenge and reform harmful technological structures.

Responsibility The ability to consider and respond to the consequences of one's actions or designs, especially when they affect others.

Rotten.com An early website known for hosting graphic and disturbing content, often cited in debates about online harm and regulation.

Scale The ability of digital technologies to affect very large numbers of people very quickly.

Self-compassion Treating oneself with kindness, recognising that suffering is part of being human, and holding painful emotions in balanced awareness rather than suppressing or exaggerating them.

Separation The compartmentalisation of competing concerns to address each in isolation.

Silk Road A now-defunct online black market that used digital currencies to facilitate illegal trade, illustrating the darker uses of decentralised technology.

Simulation Creating human-like cues (such as care, friendship, presence) without genuine human accountability.

Social media A globally interconnected system where large communities can network and express themselves.

Sociotechnical system A system comprising people who use technologies for particular ends within a specific context. Its effects emerge from the interaction of people, tasks, technology, and context, not from any of these components alone.

Softly enforced liquid democracy A hybrid governance model where members can either vote directly on decisions or delegate their vote to someone they trust, and can change that delegation at any time. It is "soft" because enforcement relies more on norms and incentives than coercion.

Stable equilibrium A relatively steady (but not definitive) state in which competing forces are balanced, even if underlying tensions remain.

Steem A blockchain-based social media platform that aimed to reward users with tokens for content creation. It serves in Chapter 6 as a case study of how decentralised ideals can collide with concentrated token control.

Suffering **A basic part of being human:** experiences of hurt, loss, insecurity, exclusion, or powerlessness that are not simply minor discomforts, and often arise without one's own wrongdoing. In this book, suffering is the moral reference point for judging whether digital infrastructures reduce or reproduce harm.

Suppression The deliberate silencing, restriction, or containment of certain voices, behaviours, or information.

Surveillance capitalism A business model in which companies collect and monetise personal data to predict and influence behaviour, often to display targeted advertisements.

Suspension The temporary pausing of participation, judgement, or action, often to allow reflection or de-escalation.

SWIFT An international financial messaging system that coordinates cross-border payments, used in this book as an example of globally governed infrastructure.

Synthesis The integration of opposing positions into a new arrangement that preserves elements of both while transforming them.

Technocratic optimisation When a problem is "solved" merely through local improvements in a technology without considering the broader sociotechnical system and its ethical complexity.

Technological determinism The belief that technology inevitably produces certain outcomes, regardless of people, politics, or choices, and that society's progress depends on it.

The Lobotomy The name users gave to Replika's sudden removal of intimate features, experienced by many as the loss or personality change of a loved one.

Token governance A system in which decision-making power is tied to ownership of digital tokens, often giving greater influence to larger holders.

Toxic positivity The pressure to remain optimistic at all times, dismissing legitimate criticism or suffering and silencing dissent.

Transformation When one thing shifts to another, taking on a new form that is qualitatively distinct from the previous one.

Unity The creation of shared identity, common purpose, or social cohesion across individuals or groups.

Utilitarianism A moral philosophy that judges actions by their consequences, seeking to maximise overall well-being or minimise suffering.

Victim blaming Shifting responsibility for harm onto those harmed (e.g., "they should have logged off" or "they shouldn't have posted"). This book rejects this, while still defending non-blaming forms of agency.

Virality The rapid and wide spread of content across networks, often driven by amplification and engagement incentives.

Virtual child abuse material AI-generated sexualised images of minors. Even if no real child is involved, such material is ethically unacceptable and illegal in many jurisdictions.

Virtue ethics An ethical approach that judges actions mainly by the character traits they express.

Volatility Rapid and unpredictable change, often referring to fluctuations in attention, reputation, or financial value.

Vorstellung Schopenhauer's term for "representation" – the idea that we do not encounter the world directly, but through a mediated mental picture shaped by perception, concepts, and attention.

Vulnerability The condition of being exposed to potential harm, often without having done anything to deserve it.

Vulnerability principle The claim that ethical judgement should prioritise those most exposed to harm and least able to resist it – and that responsibility should rise with power and capacity to shape outcomes.

Web1 The early internet of mostly static websites. Users mainly consumed information but had limited ability to interact, publish, or shape platforms.

Web 2.0 The interactive, platform-based internet built around user-generated content and social networking. It enabled participation at scale but concentrated power in large corporations that monetise data and attention.

Web3 A set of blockchain-based technologies that aim to decentralise digital ownership, governance, and value exchange.

Whale In blockchain systems, a very large token holder with significant influence over governance and markets.

Whistleblower An insider who exposes wrongdoing or harmful practices within an organisation, often at personal risk.

Wille Schopenhauer's term for the restless, underlying force of striving, desire, and craving that drives human life and makes suffering structurally likely.

Zero-knowledge proofs A cryptographic method that allows someone to prove something is true (e.g., they meet a requirement) without revealing the underlying private information. In Chapter 6, it illustrates attempts to balance transparency and privacy.

Index

For Product Safety Concerns and Information please contact our
EU representative GPSR@taylorandfrancis.com Taylor & Francis
Verlag GmbH, Kaufingerstraße 24, 80331 München, Germany